# Surviving Terrorism In America

## Ronald Laes

authorHOUSE™

*1663 LIBERTY DRIVE, SUITE 200*
*BLOOMINGTON, INDIANA 47403*
*(800) 839-8640*
*WWW.AUTHORHOUSE.COM*

First published by AuthorHouse 09/13/05

ISBN: 1-4208-7214-1 (sc)

Library of Congress Control Number: 2005906779

Printed in the United States of America
Bloomington, Indiana

This book is printed on acid-free paper.

Cover Photo compliments of Jayson Kohama, who served as a communications specialist with the National Disaster Medical System at Ground Zero after 9/11.

*To the memory of those brave men and women of the United States of America that have paid the ultimate price on battlefronts afar and on our soil in order to preserve our freedom, safety, and the American way of life.*

# Acknowledgements

Thanks to all of those selfless souls within the Emergency Response Community which I have had the pleasure of working with all these years. Special thanks to Alan Pascua, Battalion Chief of the Maui Fire Department, Edward Teixeira, Vice Director of the Hawaii State Civil Defense, and Kyle Watanabe of the Maui County Civil Defense for your teachings and trainings. My heartfelt thanks to Dr. John Mills, M.D. FACEP, Dr. Charles Mitchell, M.D. FACEP, And Sharlee Dieguez, Senior Administrator of DMAT HI-1 for your patience, trust, technical advise, and above all; your friendship.

# Table of Contents

# Introduction

It is not the intention of this book to turn the average citizen into a gun wielding "James Bond", able to disarm nuclear bombs and single-handedly conquer the forces of evil! The information within is intended to give all American citizens a better understanding of the hazards we face with terrorism (as well as non-terrorist disasters), and provide basic guidelines to aid in the decisions and actions necessary in order to minimize injuries and prevent loss of life during and after a catastrophic event.

The World Trade Center and Pentagon events of 9/11 have shown our Nation's vulnerability to terrorism as never imagined before. That day has even seemed to alter the measurement of time for Americans. We hear the words "Pre" &" Post" 9/11 daily. *Remember when time was marked with "B.C." and "A.D."?* We, as civilians can individually and collectively perform roles to help thwart future terrorist acts on American soil and minimize the impact of those events should they occur.

## New World/New Hazards

As our world becomes more technologically advanced, so do the potential hazards we face each day. With a rising world population, more and larger chemical and biological facilities and manufacturing plants are being built, and more new toxic substances are being created, transported and used in our workplaces and homes. More transportation of these substances is taking place on our highways, railroads and on aircraft. The potential for industrial accidents and terrorist acts using these substances is greater than ever before.

We, as individuals and families need to better understand how to protect ourselves from the dangers that surround us, and how to stay out of harms way. Much can be done to protect ourselves by simply taking a few moments to consider the dangers we could be exposed to within our normal daily routines. *Are there chemical plants near our homes, schools, or workplaces? What would I do, where would I go should there be an accident or act of terrorism there?* It is basic human nature to try to help others in need, but we must think before reacting these days. With the knowledge of how first responders react and what they need from you, as well as a basic understanding of explosives, radiological elements, chemical and biological agents and their identification, their symptoms, decontamination requirements and medical treatments, you can be a tremendous asset to the trained responders and enhance your own personal safety and survival.

Geo-political issues, religious and ideological rationalizations, and International Terrorism will not be discussed in great detail in this book. The contents within are not to understand *why* terrorists do what they do, but to help the reader to know *what* to look for, and *how* to react to terrorism within America.

I hope this book will reduce the fears and feelings of helplessness that so many people have conveyed to me throughout the last decade….and deliver hope for the safety of our families and Nation.

Ronald G. Laes

# Chapter One

# TERRORISM

*Terrorism*...Americans have long associated the word and concept as belonging in the Middle East, (or at least in another country). After the multiple attacks on 9/11 we have finally opened our eyes and acknowledged the fact that terrorists can and will assert themselves against targets within the United States. *But why has it taken us so long to see it?* Many Americans have been under the impression that the 9/11 incident was the first terrorist attack on American soil. In fact, there have been many prior to that day! Terrorism is unquestionably associated with Al-Qaeda, Israel, Palestine and the Irish Republicans, however a terrorist can represent a number of factions *within* the U.S, or it could be a single person with his or her own agenda. Terrorists don't always identify themselves and make demands, but when they unleash their weapons, they certainly strike terror amongst their victims. Lets not, however, lose our focus on our most hostile and determined enemy. There is no doubt that International terrorists have developed an extremely

large and powerful network. Such names as *Bin Ladin,
Al Qaeda, Zarqawi, Jihad, Taliban, Hamas, and the Holy
War* all portray the bastardization of the Muslim religion
into violent and destructive forces. The children of these
believers are taught their ideals when they are young,
and *really do believe* that sacrificing themselves for the
cause will make them "Shahid", (Martyrs) and assure
their place in the Holy Land. These are the children that
are constantly providing themselves as suicide bombers.
The Muslim terrorists do not just come from the Middle
East, but are also from France, Spain, Great Britain, the
Netherlands, and elsewhere in the world. Small "cells"
(groups), are even formed and trained in these countries,
and are all eager to join the fight and die for their cause.
The Madrid train bombing, which killed 191 people and
injured hundreds more is one example of global terrorism. Al Qaeda won the battle that day. Spain pulled her
troops out of Iraq. Do you think Spain is safe from these
terrorists now? Absolutely not!

We Americans must do everything in our power to
abstain from submitting to these terrorists, and to prevent them and their ideals from entering and destroying
our Nation.

# A Brief History of Terrorism in America

America's first taste of a terrorist was in August of
1966, when Charles Whitman, after stabbing his mother
and wife to death, killed 14 and wounded dozens of stu-
dents in a shooting spree from the University of Texas

tower that lasted 96 minutes! Whitman's agenda was not political, he just felt the world wasn't worth living in!

Politically motivated terrorism dates back as far as 1975, when a Wall Street bar was bombed by Puerto Rican Nationalists, killing 4 and injuring over 50 people. This group of terrorists, known as the F.A.L.N. ( Fuerzas Armadas de Liberaci Nacional ) spread their reign of terror in New York City and Chicago from 1974-1986. They are accountable for over 50 bombings, a dozen incindiery bombings, 4 deaths, and over 60 injured victims.

The F.A.L.N.'s agenda was to obtain independence for Puerto Rico. Many other attempts were thwarted by the F.B.I.

In February of 1993, foreign terrorists detonated a large bomb at the World Trade Center in NYC which killed 6 and injured over 1000 people. The bomb was intended to do more structural damage than it did, but served well its purpose of terror. Osama Bin-Ladin was responsible for this attack.

In 1994, Rashid Najib Baz killed one and injured 3 in a "drive by" shooting in Brooklyn. His targets were youths of the Hashid (Jewish) faith.

In April 1995, Timothy McVeigh succeeded in destroying the Arthur P. Murrah building in Oklahoma City, leaving more than 600 injured and 169 dead.

In 1996, a bomb concealed in a backpack exploded in the Centennial Olympic Park in Atlanta, Georgia. That incident killed 2 and injured over 100. The terrorist, Eric Rudolph was captured in 2003, and charged for this bombing in 2005. Rudolph, of the "Christian Identity Movement", (a White Supremacy / Anti Semitic group), who managed to evade capture by living in the wilderness for seven years was also responsible for bombing a night club and a family planning clinic in Atlanta in 1997. In 1998, another of his bombs killed one and seriously injured another at an Alabama abortion clinic. In 1997 alone there were at least 4 bombings and 6 fires at Abortion Clinics across the Nation. That is only one year of Abortion Clinic terrorism, which began in the 1980s.

From 1978 through 1995, Ted Kaczynski, a.k.a. "The Unabomber" injured 22 and killed 3 people with letter and package bombs. One bomb exploded in the cargo compartment on a commercial airliner, forcing it to make an emergency landing and causing smoke inhalation injuries to 12 passengers. Federal Agents finally apprehended him on April 3, 1996.

Those are just a few examples of terrorism in America. There are more. And there will be more. Federal, State, and County Emergency Planning Department personnel have for years been saying "It's not *if* it will happen, it's *when* ". Since 9/11 our focus has been on terrorism unleashed by Al-Qaeda and associated groups, but as past history shows, we also need to exercise diligence in identifying and stopping hate and right winged supremacy groups as well as any individual that may demonstrate the desire and or need to commit an act of terrorism.

# Defining Terrorism

The definition of terrorism has mutated throughout the years, due to changes in motives, the means that terrorism is carried out, and upon whom it is carried out. As this metamorphosis occurs, (and terrorism is occurring with frightening frequency and potency), and the *purpose* changes, Governments may change the definitions once again. In the 1930s, at a League of Nations Convention, the following definition was adopted. " *All criminal acts directed against a State and intended or calculated to create a state of terror in the minds of particular persons or a group of persons or the general public."* In the early 1980s, the United States defined terrorism as *"Premeditated, politically motivated violence perpetrated against non-combatant targets by sub-national groups or clandestine agents…"*

In 1995, by a Presidential Decision (PDD-39), the FBI definition was changed to read *"The unlawful use of force against persons or property to intimidate or coerce a government, the civilian population, or any segment thereof, in the furtherance of political or social objectives."*

The FBI has divided the definition of terrorism into 2 categories…Domestic, and International.

*Domestic Terrorism* is "the unlawful use, or threatened use, of force or violence by a group or individual based and operating entirely within the United States or its territories without foreign direction committed against persons or property to intimidate or coerce a government, the civilian population, or any segment thereof, in furtherance of political or social objectives."

*International Terrorism* is "activities that involve violent acts or acts dangerous to human life that are a violation of the criminal laws of the U.S. or of any State, or that would be a criminal violation if committed within the jurisdiction of the U.S. or of any State…or to affect the conduct of a government by assassination or kidnapping: and occur primarily outside the territorial jurisdiction of the U.S., or transcend national boundaries in terms of the means by which they are accomplished, the persons they appear intended to intimidate or coerce, or the locale in which their perpetrators operate or seek asylum."(18 U.S.C.)

Which ever definition by which terrorism is interpreted, the end result is still the same, and this method of "negotiating" must be suppressed globally as well as within the United States.

## Methodology

Certainly the single most destructive domestic incident by terrorists to date were the attacks on New York City and the pentagon in 2001.( Lets not forget the plane that went down in western Pennsylvania on that day too!) The preparations and ingenuity of the terrorists was incredible, however those attacks would (probably) not have been possible had the terrorists not been willing to sacrifice their lives in order to complete their mission. It certainly would have taken more time to prepare for, and financing the attack would have taken a great deal more money. The combination of intelligence, financing, and the will to die for their cause makes a terrorist network a deadly adversary. Since 9/11 we have tightened airports security and continue to improve sensing instruments, security procedures and passenger screening. The chances of another attack by a hijacked airliner will not be as easy in the future. Bombs aboard passenger airliners are still an option, particularly within the baggage and cargo compartments, as the task to sense explosives amongst millions of checked bags moving at an incredible speed through the system is challenging. New and advanced explosive detection technology is aggressively being developed and put into use at airports across the Nation. This technology combined with a vigorous screening of airports personnel should diminish the likelihood of on-board threats.

Car bombs have become an effective and most terrifying method of delivering devastation. With or without a suicide driver, a large car bomb can yield an enormous amount of destructive power. The size of the vehicle in conjunction with the type of explosives used are factors to consider when calculating a car bomb's power. (see Vehicle Bomb chart on page 80.) Car (and other) bombs will probably remain a favorite method of delivering terror in the years to come, due to the cost and availability of explosives and the components necessary to build them. We have all seen (thanks to the media) the devastation and chaos created by *small* car bombs continuously detonated in and around Baghdad recently. These bombs are being detonated and killing people almost every day!

Automatic weapons and other guns can be instruments to produce terror, but the chances of scenarios of massive amounts of continuous gunfire as seen in the Middle East regions happening here are remote. The use of guns in such a way would certainly be suicidal. On the other hand, snipers are an element to be reckoned with as demonstrated by 19 year old Lee Malvo in October 2002, as he left 10 dead in a 3 week killing spree in the Washington D.C. area.

A good, well trained sniper can be an exceptionally demoralizing adversary.

A well coordinated sniper campaign utilizing multiple snipers in various areas simultaneously could produce the desired results for the terrorist, however the loss of their own personnel to inflict minimum damage would tax their human resources here in America. It is unlikely that any international terrorist cells, (groups) within the U.S. would risk exposing themselves for such an attack, saving their resources for much larger and sensational or covert missions.

The possibility of such an attack, however must not be ignored!!

Let's not confine our thinking to explosives and guns. Terrorists don't. They can be quite creative inventing ways of creating terror with what is available to them. Train derailment can be accomplished without explosives causing massive carnage and loss of life, as demonstrated in the January 2005 train derailment in Glendale California where 11 died and over 100 people were injured. That was accomplished with an "unarmed" SUV left on the railroad tracks by a suicidal man that "changed his mind" and abandoned his vehicle on the tracks at the last minute.

Arson is another method of producing the desired effects for a terrorist. Fires kill and injure, they are destructive, and can hamper our industry. Fires are also relatively easy to start, and require no sophisticated equipment. For the most part, the terrorist only needs to gain access to the facility with his fuel and lighter. There has been an escalation in the frequency of industrial complex fires in 2004-05. The cause of some of these fires has been explained, but others are still under investigation. Clandestine Terrorism?

Kidnapping can be a very effective weapon as well… particularly if the media does not act responsibly! The recent videotapes of victims kidnapped in Iraq, pleading to keep their heads attached to their bodies was horrifying. Should these kidnappings/beheadings take place in America, the airing of such videos could have a profound effect on us...it will certainly impact our complacency towards terrorism!

Power grid sabotage can leave entire cities without the energy needed to provide heat, light, sanitation, etc. Imagine the gridlock of automobiles if all the streetlights were not functional in a large metropolitan area for an extended period of time. Disrupting the 911 dispatch system could also have disastrous effects on any given day, not to mention *in conjunction* with a large scale attack. Internet viruses can be a very effective weapon as well, shutting down vital emergency and government communications. Chemical and fuel spills can be brought about with hand tools alone.

There are many items and concepts that can be used to inflict terror.

The most terrifying of possible attacks are those that may come using Weapons of Mass Destruction, (WMD's) such as chemical or biological agents, radiological materials, or nuclear weapons. It is somewhat fortunate for us that these types of weapons are not easily obtained, and they are costly.

Terrorists have made many attempts to obtain the ingredients for WMD's, and do at this time have various components in their arsenals!! Unfortunately, Black Market trafficking of radioactive, chemical, and biological materials is taking place Internationally at this time.

Biological attacks are silent killers. They can decimate our population and/or our crops in a short period of time without any warnings. They can be totally invisible.

Radiological attacks can contaminate important areas of commerce, industry, transportation and Government infrastructure, leaving buildings and cities uninhabitable for *years*, and costing millions in clean up operations, not to mention the deaths and sickness that can go along with the attack.

As you can see, the possibilities and diversity of a terrorists methods are endless. Fiction writers and Hollywood have provided plenty of scenarios to consider! Actually, Law Enforcement agencies approached the movie industry after 9/11 in order to gain more insight on various types of possible incidents!

## Targets

As has been demonstrated in the last few decades, a terrorist's favorite target needs to have a certain composition. A key element in selection is *symbolism*. Naturally, the site selected will reveal the views of the terrorist.. The second ingredient is *visibility*. The more people that can observe the carnage, the better. Unfortunately, terrorists wouldn't have much success in getting their wishes granted by just blowing up a building or wrecking cars, so the last ingredient is to assure that *morbidity and mortality* are achieved. This is why the terrorist will select targets with as dense a population as possible. We have had our share of domestic terrorism for various ideals, but our greatest nemesis now is Bin Ladin and his Al Quaeda fanatics. They don't like our lifestyle, our riches, or our ideals… and they are determined to destroy all that is dear to us. Americans are their favorite target, and they have us in their sights. The events of 9/11 have shown us their desire to disrupt our economical system. Our freedom of travel has also been hampered by the events of that day. If you take the time to really think about it, you will recognize many potential targets. Our natural resources such as agriculture, forestry and water could be compromised. Attacks on our religious and prayer centers would certainly produce a desirable effect for the terrorists. Would Disneyland be a good target? You bet it would, and so would Las Vegas. Imagine the destruction of the Space Needle, the Empire State Building, or the Statue of Liberty. Just about anywhere there are large gatherings of people and significant buildings, there is a target.

It may be of interest that our TV stations are probably safe! Why? Because that is where terrorists get information!! ( I will discuss *responsible journalism* in a subsequent chapter.)

## Target Profiles

Government buildings
Financial Institutions
Mass transit means or buildings (Railroads, Subways)
Business Buildings
Health Care Facilities
Communication and Public Utility buildings
Chemical Plants
Fuel Plants and storage
Nuclear power plants
Hotels
Schools
Churches

Harbors and shipping centers
Water supplies
Cruise ships
Shopping Malls
Bridges
Border Crossings
Airports
Fire, Police, and Ambulance facilities
Significant Landmarks
Factories
Farms
Dams
*Any location where large groups of people congregate.*

As you can imagine, there are a lot of possible targets in our Country. Efforts are being made by Homeland Security and local Law Enforcement, but the task is immense and costly to constantly inspect and monitor the security of these (and other) facilities. American citizens can help by simply paying more attention to their workplaces and surroundings and report any suspicious activities to the proper authorities in a timely manner.

## Terrorist's within our borders

We the People of the United States of America need to take a close look at the security of our borders and the proliferation of illegal immigrants within our Country. The Declaration of Independence was created and signed by immigrants. Our great Nation was built by the ingenuity and ideals of immigrants from many different nationalities. Our industries and institutions that have formed the basis for our way of life were created by immigrants. *Legal* immigrants have been welcomed into, and contributed much to our society, but illegal immigration must be stopped, as with them come those that have the desire to do our people and our society harm. Recent data shows that there are over 9 million illegal aliens now living within the U.S. That is over 3% of our population! Think of the implications of these numbers continuing to multiply. Americans need to take a pro-enforcement stance against illegal aliens. This doesn't just mean stopping illegal aliens, but returning those who already exist in our country back to where they came from. If they are sincere in becoming an American citizen, then they can come in through the proper channels. We need to take a hard stance on reforming the existing illegal alien problem, and controlling the surge at our borders. We need to get tough. Controlling illegal immigration is an extremely daunting task, as our country already swells with illegals, but it must be done…and soon. Much like a sinking boat, we must stop the leaks at our borders immediately, and at the same time, "bail out" the illegal immigrants already within the Nation. Not only do we need to consider the infiltration of terrorists into our Nation, we also need to realize that tax-paying American citizens (and our economy) are suffering financially from illegal immigrants! Our doors are open to those that can legally immigrate here and become a functional, taxpaying part of *our* society.

We must question *why* a person is in our country illegally. We must also question whether or not someone *is* an illegal alien. Our bleeding hearts could lead to our spilled blood if we aren't very careful.

It is every Americans duty to question an expired student's visa or suspected illegal immigrant, and to report those individuals to the U.S. Citizenship and Immigration Services. The USCIS is a division of the Department of Homeland Security. Its predecessor was the INS. (Immigration and Naturalization Service). It really isn't that hard to do the right thing. Employers, landlords, instructors, rental agencies, etc; if suspicious, should demand to see proof of legal immigration before offering their services.

Border Guards and Customs Agents must be *extremely* careful performing their duties at all times. It was the diligence and watchful eyes of U.S. Customs Inspectors that led to the capture of Ahmed Ressam in 1999. Ressam entered the U.S. through Canada carrying bombs that were intended for the Los Angeles International Airport.

As an anecdote, 25 years ago, I worked in a manufacturing shop for 2 recent immigrants from Russia. They were still having some trouble with the English language, but knew enough to understand that I was asking them why they were here, and whether or not they had any bad intentions towards this Country. Keep in mind that this was a decade before the fall of the Soviet Union. I watched the expressions on their faces as they answered me. Humbly and seriously, they stated that they were legal immigrants, and loved America. These men and their families, after escaping Russia with only their suitcases, jumped all of the hoops of immigration, learned to speak English, and happily conformed to American society. They started their own business, employed many citizens, and became tax paying Americans.

There are those who want to come to America to better their lives and become an asset to our Country, and those who would come to destroy our people and our way of life. Let's be cautious and selective while opening our borders to foreigners.

## Terrorist Profiling and Intelligence Gathering

I am sure that I won't gain any favor from the ACLU, but I am in favor of profiling people in the effort to safeguard our Nation from terrorists. Profiling is not discriminating. It is a means of protecting our Nation. Profiling is "the method of determining the most likely types of individuals that would conduct themselves in a certain manner". This profiling process involves the consideration of ethnic background, age, sex, mannerisms, and other means. Police departments and the FBI have used profiling techniques for years, resulting in the arrests of many criminals. Granted, profiling isn't always 100% foolproof, but it does help narrow our searches for possible perpetrators. If you were ever a landlord or employer, I am sure you have used some means of profiling in your selection of employees or renters! Right?...I would. No way would I have a weirdo wearing a t-shirt proclaiming "Rape Now" living in the same complex as my 12 year old daughter! I do admire the efforts of the newly formed TSA in its efforts to "randomly select" travelers for supplemental inspection prior to flight take-off shortly after 9/11, but remain bewildered by its results. More profiling should have been taking place at this time in addition to the random inspections.

I was flying from Maui to Seattle a year after 9/11. I watched as a 7 year old girl was pulled from the line and extensively searched while 2 males, approximately 30 years old, of Mid Eastern persuasion, and looking anxious, stood by and were granted their boarding passes without further question. (Their profiles matched the perpetrators of the 9/11 incidents). Profiling is one effective way of identifying a possible perpetrator. Let's use it as a tool for fighting terrorism on our soil. Not only should our Law Enforcement agencies use it, but we as individuals should also keep it in mind as we go through our daily routines!

Many people fear the loss of privacy since the inception of the "Patriot Act" of 2001. I don't believe that *Patriots* should be concerned… The Patriot Act,(HR 3162) is "The uniting and strengthening America by providing appropriate tools required to intercept and obstruct terrorism." We must be allowed to use a different set of rules with the terrorist than we do in normal law enforcement. The terrorists use their own rules, and it's obvious that they don't observe the Geneva Convention! Through the Patriot Act, Counter Terrorism specialists can now use enhanced surveillance procedures for suspected terrorists, employ better Immigration and Visa integrity and security, and provide better border protection. The ability for Law Enforcement Agencies to share information and use improved intelligence methods in conjunction with strengthening the criminal laws against Terrorism should greatly improve our ability to keep Terrorists from striking within our Nation again.

Let's not let Civil Liberties interfere with our quest to catch terrorists. Profiling and investigating individuals that match the profiles of terrorists shouldn't  irritate anyone.….

Unless they have something to hide!!

## Chapter Two

# EMERGENCY PREPAREDNESS

Disaster can strike at any time. Hurricanes, floods, tornadoes, high winds, earthquakes, industrial accidents and acts of terrorism can greatly affect you and your family in the blink of an eye. You may be evacuated or restricted from being in your home, or possibly confined to or quarantined within your home for days...or longer. During a major event where either of these scenarios exists, it may take hours or days before you can receive help of any kind. You may remember news footage of the Florida hurricanes of 2004, where semi trucks carrying water and ice to the victims of the hurricanes stood idling, unable to release their cargo for *days* due to the inability to locate government approved staging areas to distribute these supplies. This was a frustrating situation for both the victims and the rescue workers, but it happened, and similar incidents are likely to happen again. *All citizens should be able to be self-sufficient for a minimum of 3 days.*

Emergency personnel such as Fire Fighters, Police and Emergency Medical Services are often taxed to their limits during their regular, non-disaster related duties. When a significant disaster strikes, they will need all the help that they can get! One thing to keep in mind is that the help they need is often for civilians to stay out of their way!

With a little thought and planning, you can give you and your family a much better chance at survival (and comfort).The best place to start preparing is right at home at the dinner table or your living room. Share your thoughts of preparedness with your family members and get them involved. A good start is to "inspect" your house and property as a group, and see if you already have disasters waiting to happen within your own home.  You might be surprised at what you find. Check for fire hazards such as frayed electrical cords or extension cords that are worn or have too many appliances plugged into them. Replace them with appliance cords or power strips with fuses. Check the batteries in smoke detectors, and test them. Locate the power panel, water and gas shutoff devises and make sure they work. Instruct those old enough on how to shut them off if the need arises in your absence. In short, make sure your home is not a disaster waiting to happen.

You should also make checklists (such as to the locations and operations of shutoffs) and keep them posted behind a closet or cabinet door. These checklists should also contain emergency phone numbers for a poison center, the family doctor, the FBI, local Emergency Management Departments and nearby police and fire stations.

All jurisdictions throughout the U.S. have some sort of Emergency Management Department or Agency. For many small towns, it is the Fire Department. Larger towns/Counties have established Emergency Operations Centers that are manned during working hours, and open 24 hours a day during an impending or existing crisis. If you have access to the internet, try searching for the Emergency Agency in your area. You might be surprised at the amount of information that is available to you, and at the amount of planning that is taking place to protect the population, its homes, schools, workplaces and the environment. Most local Emergency Departments have excellent websites that list hazards, evacuation routes, preparedness guidelines and more. Should you not have internet capabilities, call them. The emergency coordinators are always willing to share pertinent information with their population. You may even want to request a family "field trip" to a center. It is often more effective influencing children by hearing what "professionals" have to say. They will also have pamphlets and visual aids. However you do it, it is important that you get the whole family involved and "on the same page" while making your plans. Next, put the family in the car on a quiet Sunday morning and drive around your neighborhood and look for potential hazards. Include areas around your schools, workplaces and shopping malls. You might want to take a map and note taking supplies with you and list any potential industrial hazards or terrorist targets. Pay attention to hazardous material warning labels and signs. This procedure is called a *"Risk and Vulnerability Assessment"*. You may be surprised at what you find! Try to locate factories, chemical plants, railways, and evacuation routes. Your local Emergency Department can also provide you with Risk and vulnerability assessments upon your request. The next thing to consider is *prevail-*

*ing wind direction.* You will need to use the normal wind direction in any given area when plotting out evacuation routes to and from your home. This is very important, as during and after an industrial accident or terrorist event, you need to be *upwind*! Think of standing around a barbeque…Do you stand downwind and choke on the smoke? The danger of industrial accidents and terrorist attacks is that *you may not be able to see, smell or hear the hazards until it is too late.* Remember when making your plans to select alternate routes should the wind direction change! Should you and your family map out these routes and predetermine the actions and routes you will take, you will have done much to alleviate anxiety and minimize your risk of becoming a victim during a major event. Knowing the most likely actions your family has prepared for can also help you locate missing family members should the need arise.

## The Life Cycle of a Disaster

When Emergency Planners perform risk and vulnerability assessments, they not only need to evaluate the actions that will need to take place at the time of the incident, but those before and after as well. While making your family disaster plans, you should also use the same model, which is to include all 4 phases of the disaster cycle.

A disaster has 4 phases to be managed, and can best be visualized as having a circular relationship to each other. The 4 phases are: Mitigation, Preparedness, Response, and Recovery. After you study these phases, you should be able to visualize their circular relationship.

Phase 1: ***Mitigation.*** Mitigation refers to activities which eliminate or reduce the chance of the occurrence or effects of a disaster. (When you checked your house and property and replaced electrical cords, checked for gas leaks, etc; You were performing tasks associated with mitigation.)

Phase 2: ***Preparedness.*** Preparedness is planning how to respond in case an emergency or disaster occurs and working to increase your resources and plans to be able to respond effectively. (When you evaluated evacuation routes you were in the preparedness stage). Note that the definition includes the word "resources". Resources include basic needs to be self sufficient, such as food, water, first aid, etc. I will go into greater detail later in this chapter.

Phase 3: ***Response.*** Response activities occur during and immediately following a disaster. These activities are based upon the preparations made in phase 2, and should help you to avoid being a victim, help you if you or a family member are a victim, and reduce the likelihood of secondary damage.

Being able to respond safely and in as orderly a fashion as possible when disaster strikes depends entirely upon the amount of effort you put upon preparedness.

Phase 4: ***Recovery.*** Recovery is the final phase of the emergency management cycle. The recovery phase will continue until all systems are returned to normal, or near normal. *Short term recovery* restores vital life support systems.(Such as reuniting family members, restoring power, water, safety, and receiving medical attention.)

*Long term recovery* may go from months to years, depending on the severity of the incident. This can include financing and repairing damaged property, as well as psychological issues that may have impacted a family member or members.

While you are in the preparedness stage, you should pay attention to the issues involved with recovery and try to minimize the problems that can be associated with recovery such as the protection of legal and financial documents, birth certificates, and insurance policies. You may choose a fire safe, and/or store duplicates of these documents in a locale other than your home. Storing duplicates of these documents in a safe locale is an excellent plan for those living in hurricane, tornado, flood and Tsunami areas.

Now you should be able to see how the 4 phases form a circle. After recovering from a disaster, you will need to evaluate what went wrong (and right) and begin mitigations to improve your preparedness, thereby improving a future response. These principles work well for most natural and man-made disasters, and should prove helpful when near a terrorist attack as well.

## Basic Family Disaster Planning Guidelines

**1.** Conduct a *Risk and Vulnerability Assessment* of your area.

**2.** Research your local Emergency Department's existing plans and protocols, such as established evacuation routes, shelters, and emergency warning systems. You should know what the warning sirens sound like, and what actions to take for the different warning tones, should there be more than one type for your area.

**3.** Find out what a.m. radio stations are used to notify the public when disaster strikes. Program the station into your car and home stereos.

**4.** Consider your Pets! This is important, as some shelters and food kitchens will not permit pets.

**5.** Consider family members with *special needs*. People that are blind, elderly, hard of hearing, or will require mobility assistance and special medications will need special care.

**6.** Select 3 different places to meet following a disaster. One should be at home, one should be just outside your neighborhood ( a store or restaurant, etc), and one should be in another town which is usually *upwind* from your town. All family members should keep the addresses and phone numbers of these locations with them.)

**7.** Develop a *communication plan*. When family members are separated, they need a way to contact each other.

Remember that during major disasters, cell-phones and regular phone service may be interrupted or overwhelmed, and the lines will be busy. It is also a good idea to establish an "out of town" emergency contact. This can be a friend or relative. It is common that phone lines may be overwhelmed around the disaster site, but long distance calls *can* often be made. Remember to include the Red Cross in your communication plan, as they provide family services and keep a register of survivors, victims, and their locations.

**8.** Complete checklists and make sure everyone understands them and knows where they are. Make emergency phone number cards (wallet size) and have everyone keep one with them at all times.

**9.** Practice your plan periodically! This does not have to be a full scale scenario. Question and test each other. Make sure everyone has their phone number cards. Mark your calendar to review and enhance your plans every 6 months.

You should also remember these guidelines while on vacations or out of town on business. Take the time before going on vacation to explore the natural hazards in the area so you can have a better idea of how you would respond should a disaster happen, and what kind of emergency items to take with you. When arriving at your destination, take the time to familiarize yourselves with fire escapes, fire extinguishers, etc. Establish pre-determined meeting places. Take a small first aid kit with you. Think about what you would need if you were stuck in a disaster area, not knowing anyone that could help you.

## Basic Disaster Supply List

Basic supplies can be put into 3 different categories.
1) Supplies that you keep in your vehicle.
2) Supplies that are "stockpiled" at your home.
3) Supplies that you will need to evacuate or shelter with.

Common sense should be used when determining which items should be included in your cache. A sleeping bag would not be a high priority item in Southern California or Hawaii, but sure would be if you lived in a colder region like Colorado! When selecting food supplies to store, don't buy brands that you wouldn't normally want to eat just because they are inexpensive. You will want to rotate these foods into your regular meals periodically before they expire.

Acquiring the necessary supplies should not put you in debt. Prioritize the items you feel are the most important and begin collecting. Please note that the supply lists are basic guidelines and can be changed to meet your needs.

## 1. Vehicle:

- ➢ A change of clothes
- ➢ Shoes or Boots
- ➢ Raincoat or poncho
- ➢ Blanket or sleeping bag
- ➢ Flashlight, AM Radio and spare batteries
- ➢ matches or lighter
- ➢ Towel
- ➢ *Drinking water and 1 gallon or more rinsing water
- ➢ Water purification tablets
- ➢ 1 pint Bleach
- ➢ Insect Repellant
- ➢ First aid kit (small)
- ➢ 2 "N-95" respirator masks
- ➢ 3 day supply of prescription medications (*in glove compartment or on person so you can rotate or change them when necessary)*
- ➢ Spare eyeglasses
- ➢ Knife
- ➢ Signal flare or strobe
- ➢ Food ( non- perishable, canned or otherwise.)
- ➢ Paper Towels
- ➢ 2 large plastic garbage bags

Pack the items in a small backpack or duffel bag. Store the bag inside a garbage bag to protect from moisture. This basic kit doesn't take up much room, and can make all the difference in comfort and survival should you be stranded away from home for an extended period of time.

## 2. Home:

There are several reasons you may be required to shelter at home, including a biological or chemical attack. You should be prepared to totally isolate your house. This isolation should include entrances and exits. Assuring that you have good, airtight weather stripping on all doors and windows will help. If not, you can use duct tape and/or plastic sheeting over any openings. When isolating your house, consider all the ways that vapors can get into your house. Seal off fireplaces and ducting vents that lead to the outside. If you have an air conditioner, have "hepa" filters installed. These filters will stop smaller particles from entering the system than ordinary filters. In addition to isolating the whole house as best you can, you should select and thoroughly seal one or two rooms to use as your *primary isolation habitation* area. Ideally this would be an adjoining kitchen and bath. These 2 specific rooms can provide sanitation, food preparation and water, as well as sleeping areas…Are you ready to create such an area in your home?

**Water**…The more, the better!

Bottled water can be purchased, or you can store tap water in clean plastic containers or jerricans. Avoid using glass containers if possible. Rotate water supplies every 6 months. A 5 day supply of water should be adequate. Assume one gallon per day per person. Also remember that this supply is not just for drinking, but for cooking, and sanitation.

**Food.**

Select foods that require no refrigeration, minimum preparation and little or no cooking. As with the water, keep 5 days supply of food. Don't forget special diets and pet foods! It is also suggested that parents remember some "fun" foods for young children. Canned juices, coffee, tea, and snack foods can add a little pleasure for all.

**First Aid Kit.**

*Every family should have a first aid kit prepared at all times!* This kit should not "supplement" the normal items that you use every day, such as band-aids and antibiotic ointments. A good emergency first aid kit should be complete, and packed in a small bag or pack, ready to grab at a moments notice.

- Roller gauze bandage (4 rolls)
- Compress bandages
- Antiseptic solution
- Anti-biotic ointment
- Band-aids, 1" strips (1 box)
- Band-aids, large (1 box)

- Triangular bandage (3)
- Gauze sponges (10)
- Anti-bacterial soap
- Povodine/Iodine scrub solution
- Thermometer (oral)
- Alcohol
- Scissors
- Tweezers
- Aspirin or other pain relievers
- Antacid
- Activated charcoal (poison antidote)
- Eyewash
- Waterproof tape (2 rolls)
- Bleach ( for water purification and sanitation)
- Water purification tablets
- "N-95" respirators
- Paper towels
- *Prescription Medications* can be quickly added.

**Evacuation:**

Evacuation supplies should be much like those carried in your vehicle, but the list should be enhanced as you may be away from home for more than a few days.

Remember that you may be in austere conditions or living amongst many other people in a shelter. While packing these supplies, it is wise to use backpacks or duffle bags. They should be easy to carry and as light as possible.

- ➢ 3 gallons of water per person
- ➢ 3 day supply of food per person ( Pet food??)
- ➢ A pint of bleach (water purification, sanitation)

➢ Water purification tablets
➢ Insect Repellant
➢ Dish soap (small bottle) / Handsoap
➢ can opener
➢ Sturdy plastic spoons, forks
➢ Unbreakable coffee cup for each member
➢ mess-kit
➢ Paper towels & toilet paper
➢ N-95 Dust mask respirators (6 per person)
➢ Large Safety pins (to be used as clothes pins, etc)
➢ String (heavy) or ¼" nylon rope (50 ft.)
➢ Plastic garbage bags (10)
➢ Playing cards, paperback book, writing tablet/ pen
➢ Identification
➢ Cash and credit cards
➢ *Prescription medicines*

As previously stated, these lists are only *guidelines* to get you started. You will need to configure your supplies in accordance to your geographical region and specific family needs. As with the first aid kit, your evacuation bag should be at least partially packed. A checklist of items that need to be added to the bag should be attached to the outside of the bag so items are not forgotten. Empty water bottles can be included, with a note to fill them with fresh water before leaving. If, after preparing the family evacuation bag, you should find it too heavy or cumbersome for one individual to carry, consider using 2 smaller bags. Along with the first aid and evacuation kit, there should also be a backpack for each family member.

This is for clothing, water, personal hygiene supplies, and medications. You should not rely on wheeled luggage for evacuation bags, as the terrain you travel over may not be suitable for them! Rely upon what you can *carry!*

If you haven't already made similar plans and stockpiles, don't procrastinate. Set yourself a goal date and get started. Locate bags and backpacks. Conduct a "practice" evacuation with the entire family, where all bags are packed and tested for weight and functionality. Take the time to plan for and prepare and stock supplies that would be needed to *quickly* fabricate a primary isolation habitat area.

*Remember that you should be able to remain self sufficient for a minimum of 72 hours!!*

**Water Purification**

Lastly, I feel the issue of being able to have safe drinking water is vital to any preparedness plan. Our bodies need water to survive. During or after a disaster, you may not be able to get drinking water from a tap…or a store. There are 3 basic ways for a person without a sophisticated water purification system to provide potable water.

**1. Purification Tablets:** Iodine tablets can be purchased at sports stores. A bottle of 50 tablets is small, and can treat 25 canteens of questionable water. These should be added to your emergency evacuation supplies, as they don't require anything more than a water bottle and water. The tablet is placed in the water and shaken. Most water purification tablets provide potable water in about 30 minutes.

**2. Boiling:** Boiling is by all means the safest method of treating water for drinking. Begin by straining the water to remove large impurities. Boil the water vigorously for over a minute. Let the water cool before drinking.

**3. Bleach:** Regular unscented household bleach is 5.25% Sodium Hypochlorite. This will kill most common bacteria. Add 16 drops per gallon to previously strained water. Stir or shake. Let stand for 30 minutes. The water should have a faint smell of chlorine. If not, the contaminants in that particular water supply are greater than the chlorines ability to treat. If so, the water should be discarded and another source found.

There are other means of obtaining drinking water, such as distillation and creatively searching sources such as water heaters and hose bibs or home water pipes, but the 3 methods suggested are relatively fast and require little equipment.

# Emergency Related Internet Resource Links

**American Red Cross**
http://www.redcross.org/

**Ready.Gov (Department of Homeland Security)**
http://www.ready.gov/

**National Disaster Education Coalition**
http://www.disastereducation.org/default.html

**Prepare.Org**
http.www.prepare.org/

**FEMA**
http://www.fema.gov/

**National Organization on Disability**
http://www.nod.org/emergency/index.cfm

**American Society for the Prevention of Cruelty to Animals**
http://www.aspca.org/site/PageServer?pagename=emergency

# Chapter Three

# WEAPONS OF MASS DESTRUCTION

## Definition

Weapons of Mass Destruction ( acronym WMD) are weapons that are designed to kill large numbers of people, civilians and military personnel alike. WMDs, while causing mass destruction and death, also have a severe psychological impact on the population.

The military definition is "Weapons that are capable of a high order of destruction and/or of being used to destroy large numbers of people. WMDs can be high explosives or nuclear, biological, chemical, and radiological weapons, but exclude the means of transporting or propelling the weapon where such means is a separable and divisible part of the weapon" (D.O.D Joint publication 1-02)

Presently, United States Civilian Defense agencies have adopted the term CBRNE (pronounced "see

Bernie") This definition includes **C**hemical, **B**iological, **R**adiological, **N**uclear, and **E**xplosives.

Weapons of Mass Destruction have been around and in use for centuries. Only recently have we been made aware of the term WMD. Most Americans have ignored the fact of their existence until brought to light recently in our search for them in Iraq. The truth is, many countries presently store numerous varieties of WMD...including America. The phrase WMD was adopted in 1937 to describe large quantities of conventional bombs. Throughout the Cold War which began shortly after World War II and until the fall of the Soviet Union in 1991, our military termed these weapons as NBC, or "nuclear, biological and chemical" weapons. Most focus during that period of time was on the possibility of nuclear attacks between nations. Both Russia and the United States developed and readied thousands of nuclearwarheads and were prepared to deploy them should aggression be imminent. Those were scary times for U.S. citizens. Baby Boomers and older generations remember the scramble to build underground bomb shelters and stockpile food and supplies for survival should the "unimaginable" happen. The unimaginable came very close during the Bay of Pigs era in 1962, when the U.S. attempted to reverse the nuclear warheads Russia was placing in Cuba. During the Cold War era, Russia and America (as well as other countries) were simultaneously manufacturing and stockpiling chemical and biological weapons. The proliferation of these WMDs was enormous. Thousands of *tons* of warheads were produced. Many of these stockpiles have been destroyed, but many still exist. America's stockpiles are slowly but surely being destroyed. Most of our stockpiles exist close to cities with significant populations. In August of 2003, thousands of residents of Anniston Alabama were issued hood style gas masks in the event that an accident in the destruction of large amounts of nerve gas warheads being stored there should occur. Progress in the destruction (incineration) of these aging weapons is proceeding, although hampered by concern for the public's safety. Many people want the weapons transported to remote sites for disposal, but that also presents a hazard. Transportation of these materials can be an extremely dangerous task. The method and security of transporting these weapons poses many problems. Train or truck transportation is susceptible to accidents or terrorist attacks as they move along the highways or railways through our cities. For the most part, our existing stockpiles are being heavily guarded from theft and attack. Our greatest fear is the disposition of those weapons that existed (exist) in the former Soviet Union. Not only do we fear the disappearance or misuse of the weapons, but identities, locations, and intentions of the scientists and specialists that developed the weapons. Tens of thousands of scientists and technicians were displaced and out of work since the demise of the Soviet Union in December of 1991. Many of them had enjoyed a comfortable salary for years, and are now without jobs. The lure of money has tempted some to emigrate to other countries in search of employment. Our greatest fear is that these brilliant scientists may end up working for rogue states and terrorist organizations such as Al-Qaeda. Previously we dismissed the idea that terrorists would have the means to acquire the materials or technicians to construct and deliver a nuclear devise, but it is now highly possible that could happen. The financial resources of Bin Laden and the Taliban generates concerns regarding the ability to deploy such weapons.

An undetermined amount of nuclear, chemical, and biological weapons are missing from Soviet stockpiles. The Soviet Union's weapons program was so secret, vigorous, and dispersed, that even top officials don't have accurate accountings of the weapons. In 1997, Russia's former Chief of National Security stated that they may have lost up to 100 one kiloton "suitcase" nuclear bombs. Much is being done Internationally to find these missing weapons and eliminate the possibility of them falling into the hands of terrorists. Hundreds of tons of weapons grade plutonium and uranium have been located and secured, thousands of warheads, and hundreds of ICBM silos have been deactivated and or destroyed in Russia since 1992. The Nunn-Lugar Program deactivated more than 300 nuclear warheads in 2004 alone, but there exists the greatest possibility that other weapons are either en-route to or in the hands of the wrong people. It is of dire importance that radicals and terrorists are denied the technology or the means to deploy the deadliest of all Weapons of Mass Destruction; the chemical, biological, nuclear and radiological elements.

## History of Chemicals in Warfare

The earliest recorded use of chemicals as a weapon was during the Peloponnesian War in 423 B.C., where smoke was directed at the enemy through hollow beams. This irritated their eyes and interfered with their vision. In the 6th century, "Greek Fire" was invented. This was a substance made with rosin, sulfur, pitch, lime and saltpeter. When ignited and deployed upon the enemy, it would stick on the skin and burn persistently. Greek fire was the precursor to Napalm. During the American Civil War, Chlorine was considered for use as a weapon, but dismissed. The first large scale use of a chemical weapon was during WW I. The German military released over 100 tons of chlorine from thousands of pressurized canisters in Belgium in April 1915. 800 Allied soldiers died, and over 2500 were incapacitated in the attack. Phosgene, which was far more deadly than chlorine was also used late in 1915, and in 1917 Germany unleashed mustard gas on the Allies. By wars end, over 90,000 were dead and over 1 million casualties were sustained. During World War II, Germany was developing nerve agents, however no chemical warfare was used in the war. It is interesting that the designations we give to nerve agents all begin with a letter "G" (for German). After the war, Allies discovered these chemical agents. Tabun, being the first found was named GA. Sarin was the second, therefore named GB. During the early Cold War era, the race began, and research and development of chemical weapons was put into high gear by both Russia and the United States. The Vietnam War brought into use "Agent Orange, Purple, White, and Blue" for use as defoliants in order to eliminate the enemies forests and jungles where they hid. Unfortunately our soldiers were also exposed to the chemical, and many suffer from their effects to this day.

# Chemical Classifications

There are 2 ways in which chemical weapons are classified, persistency and type.

## Persistency

A *Persistent* agent is an agent that will continue to present a danger for a considerable amount of time after it is released. These agents take time to vaporize. They tend to be oily or sticky.

A *Non-Persistent* agent disperses or vaporizes quickly after it is released, and presents an immediate but short lived hazard.

Non persistent agents *can* be made persistent by "thickening". This is done by combining the agent with polymers. This increases the agent's viscosity.

## Types

Nerve Agents (neurotoxins)
Blood Agents (cyanides)
Blister Agents (vesicants)
Riot Control Agents (lacrimating)

The types of agents listed above are the most common agents that have been developed for military use. Terrorists may find these difficult to obtain or deploy, and find other chemicals that would be effective for their cause. Some examples that are commonly used in industry are anhydrous ammonia, hydrogen fluoride, sulfur dioxide, and chlorine. Any chemical that causes pain, discomfort or death can be used.

*Not all chemical agents will "act" the same. Some may show serious effects immediately, and others may take minutes to hours to begin causing irritation and other symptoms associated with the particular agent.*

Some common symptoms are *shortness of breath, eye & nose irritation, skin irritation, and upper airway irritation. Vomiting* can also be a sign of chemical exposure. Should you happen upon a scene where multiple persons (animals and birds included) are displaying similar symptoms, you are probably witnessing a chemical attack (or release).

# How a Chemical Weapon is Delivered

How a chemical weapon is deployed depends on what means are available. Military delivery systems rely upon missiles, projectiles, mines and bombs. If the terrorist hasn't procured military delivery systems, he could devise many ways to deliver the agents. The most effective system would be to turn the agent into an aerosol. By reducing the agent to microscopic sized particles, (5 microns would be adequate) the particles, after release would tend to "float" above the ground better. The ultimate goal of the terrorist will be to get as many particles into as many person's *lungs* as possible. Crop dusting airplanes would be a great way to disperse chemicals over a highly populated area. A garden variety hand-pump insecticide sprayer in the hands of a fast runner or from a balcony would be quite effective as well.  As you can see, the possibilities are as endless as the equipment and chemicals commercially available to the terrorist.Weather conditions can dictate the successfulness of a chemical weapon. Winds can either work for or against the terrorist, blowing the agent over the intended victims, or hopefully back at the perpetrator! Temperature, rain, and humidity are also factors that can help or hinder an attack.

*It is important to remember that the most devastating route of entry will be through your lungs, but exposure to the eyes and skin can be irritating and deadly as well.* **Remember to move and stay up-wind!**

## NERVE AGENTS

Nerve agents are *toxins*. They were developed by the militaries to be extremely toxic to the enemy, but to "break down" rapidly so the attacking forces can move into the area within days. Nerve agents are organophosphates, much like insecticides such as Malathion, Diazanon, and Parathion. Military nerve agents are odorless, non-military agents do have an odor.

The routes of entry of all nerve agents are primarily through *inhalation and skin absorption.* All nerve agent victims will require *decontamination.* The amount of a nerve agent required for a lethal dose can best be visualized by looking at a penny. On the back of a penny is an imprint of the Lincoln Memorial. A lethal dose is the amount of agent that can fit between 2 of the columns!

## Signs and Symptoms of Nerve Agents

*Dead birds and animals present are indicators that a nerve agent has been released.*

The initial symptoms are referred to as *slud*. Slud means **S**alivation, **L**acrimation, **U**rination, and **D**efecation. The victim will be in agony. Advanced effects are *seizures, convulsions,* and *death.*

Mild exposures will present symptoms within minutes. A large dose will present symptoms within *seconds*.

## Medical Treatment

*Immediate injection of* 2-pam chloride, atropine, diazepam. Establish oxygen delivery. Decontamination.

## SARIN (GB)
Isopropyl methylphosphonfluoridate
Boiling point: 297 F   Vapor density: 4.86

**Symptoms onset:** Vapor: seconds/ Liquid: minutes to hours.
**Exposure:** Liquid or vapors can be fatal. Clothing releases agent for about 30 minutes after contact with vapor.
**Odor:** Fruity
**Persistency:** Moderate/ less than 24 hours
**Isolation distance:** 2000 feet
**Evacuation distance:** 5.5 miles

## TABUN (GA)
*N-dimethylphosphoramidocyanidate*
Boiling point: 475 F   Vapor density: 5.6

**Symptoms onset:** Vapor: seconds/Liquid: minutes to hours

**Exposure:** Liquid or vapors can be fatal. Clothing releases vapors for about 30 minutes after contact with vapor
**Odor:** Fruity
**Persistency:** Moderate/ less than 24 hours
**Isolation distance:** 2000 feet
**Evacuation distance:** 5.5 miles

## SOMAN (GD)

*Pinacolyl methylphosphonofluoridate*
Boiling point: 225 F   Vapor density: 6.33

**Symptoms onset:** Vapor: seconds/ Liquid: minutes to hours
**Exposure:** Liquid vapors can be fatal.
**Odor:** Camphor
**Persistency:** Moderate/ less than 24 hours
**Isolation distance:** 2000 feet
**Evacuation distance:** 5.5 miles

## VX

*O-ethyl S-(2-iisopropylaminoethyl) methylphosphonothiolate*
Boiling point: 568 F   Vapor density: 9.2

**Symptoms onset:** Immediate
**Odor:** Sulfur
**Persistency:** Moderate/ less than 24 hours
**Isolation distance:** 2000 feet
**Evacuation distance:** 5.5 miles

Nerve agents are on the top of the list of concerns when first responders (EMS and Fire Fighters) contemplate and plan for their response to a chemical attack. This is partly because of the swift and deadly effects of the toxin, and in the event of a large number of victims, the antidotes may not be available in sufficient quantities. The antidote for nerve agents is called a "Mark 1 auto-injector". Contrary to the scene in "The Rock", where Nicholas Cage plunges an enormous needle into his heart, the auto-injector has smaller needles, and is injected into the thigh muscle. Also contrary to that movie are the initial effects of the gas. It does *not* melt the victim as it works! First responders, if wearing their personal protective equipment, can administer the antidotes before the victim has been decontaminated, but this initial treatment must take place *immediately.* It must be remembered that most EMS policies will *not* permit patients in the ambulance without being thoroughly decontaminated. Should you be unfortunate enough to have been in the area during or shortly after a nerve agent attack, it is imperative that your clothes be removed and you be *decontaminated* as quickly as possible *whether you show adverse symptoms or not.*

# BLOOD AGENTS

Blood agents are members of the cyanide family. Typical military agents are Cyanogen Chloride and Hydrogen Cyanide. Both are colorless, highly volatile liquids. Other forms of cyanide are Potassium and Sodium Cyanide. These are solids or powders that can either be released as powders or dissolved in water. They are frequently used in industry and relatively easy to acquire.

All cyanides may be absorbed in the lungs and through the skin, eyes and mucous membranes. Blood agents work by interfering with oxygen utilization in the bodies cells. *Respiratory failure is the usual cause of death, but heart and brain cells are initially affected.*

### Signs and symptoms of blood agents

Weakness in the legs
Nausea
Vertigo
Headache
Vomiting
Convulsions
Increased respirations
Cardiac symptoms

Exposure to Cyanogen Chloride will also produce intense and immediate irritation to the nose, throat, and eyes.

**Medical Treatment**

Decontamination. Respiration by ventilator. Amyl Nitrite. Intravenous Sodium Nitrite. Sodium Thiosulphate.

# Hydrogen Cyanide (AC)
*HCN*
Boiling point: 19 F   Vapor density: 0.93
**Symptoms onset:** Within seconds
**Odor:** Bitter Almonds or Peach kernels
**Persistency:** Vaporizes quickly. Gas rises and dissipates.
**Isolation distance:** 200-600 feet
**Evacuation distance:** ½ mile

# Cyanogen Chloride (CK)
*CNCL*
Boiling point: 55 F   Vapor density: 2.1
**Symptoms onset:** Within seconds
**Odor:** Faint bitter almonds
**Persistency:** As it is twice as heavy as air, it lingers in lower areas, will remain for longer periods of time.
**Isolation distance:** 200-600 feet
**Evacuation distance:** ½ mile

Due to the nature of cyanides as a vapor, decontamination is not necessarily required, however if it is delivered as a liquid or powder, decon should be performed as quickly as possible.

Patients that survive an initial exposure should be hospitalized and closely monitored, as further clinical problems could arise.

# BLISTER AGENTS

There are 3 basic Blister Agents. Mustard, Lewisite, and Phosgene. Each has its own specific characteristics, but all are *vesicants*. The properties of vesicants are the blistering effect like poison ivy produces. The blister agents were designed to produce mass casualties, they were not designed to kill, although exposure to these agents *can* prove fatal. It is blister agents that Saddam Hussein purportedly used on the Iranians in the 1980s. The blister agents cause large and extremely painful blisters upon contact with the skin. These agents do present an inhalation hazard due to vaporization, but most commonly effected areas of the body are the eyes and skin, particularly areas that have high concentrations of *perspiration.* The most lethal aspect of blister agents is when they are inhaled into the airway and lungs. The blistering effect can close the airway, suffocating the victim. If it is quickly recognized by medical responders that a victim has inhaled mustard gas, an airway must be inserted to prevent the blisters from closing the passage of air to the lungs.

*It must be noted that symptoms from mustard gas exposure may not be present for hours after contamination.* Should you suspect exposure to a blister agent, decontaminate as quickly and thoroughly as possible! Should you begin experiencing *swelling or burning sensations in the throat or lungs* after a suspected blister agent attack, seek *immediate medical attention in order to establish an airway.* Long term problems and death can occur due to infections caused by the rupturing of the blisters within the throat and lungs.

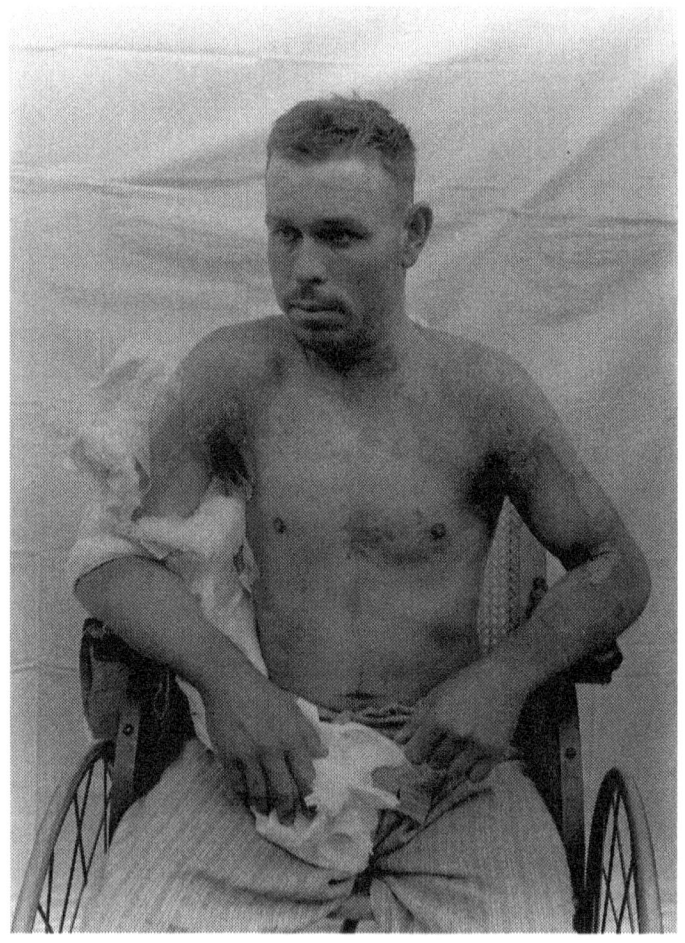

**Mustard Gas blisters. France, World War I.**

Photo courtesy of the National Archives

## Nitrogen Mustard (HD)

There are 3 Nitrogen Mustards, NH-1, NH-2, and NH-3. All 3 toxic Mustards share the same vesicant properties and symptoms. The color of the agents ranges from colorless to a pale yellow. Each has its own unique odor.

**NH-1:** Odor: Fishy, musty

**NH-2:** Odor- Fruity, soapy

**NH-3:** Odor- Odorless (May have a butter almond scent)

**Symptoms onset:** *2-6 hours* after exposure.

**Persistency:** HIGH, oily substance.

**Isolation distance:** 2000 feet

**Evacuation distance:** 5 miles

**Symptoms:** *Blistering of the skin, burning sensation of the skin, skin lesions, nasal or lung irritation, runny nose, coughing, shortness of breath, irritation or burning of the eyes.*

## Lewisite (L)

Lewisite, a vesicant, also contains *arsenic*. The presence of arsenic compounds the effects of this blister agent, as it is a poison and can cause dangerous low blood pressure as well as stomach problems. Lewisite is a clear, oily substance in its pure form, but can have an amber to black color depending on its purity.(Amber being the purest.)

**Odor:** Geranium

**Symptoms onset:** Immediate

**Persistency:** HIGH, Oily substance.

**Isolation distance:** 2000 feet

**Evacuation distance:** 5 ½ miles

**Symptoms**

Skin pain and irritation within a minute or less. Redness of affected skin areas within ½ hour. Blisters on the skin can appear within hours. Eye pain and irritation, runny or bloody nose, sneezing, coughing, diarrhea, vomiting, shortness of breath, and nausea.

## Phosgene (CX)

Phosgene is widely used in industry. Hundreds of thousands of tons are produced each year. In 1984, at a chemical plant in Bhopal, India, a leak occurred that released 25 tons of the vapor, killing 200 and injuring over a quarter of a million people. That was industrial phosgene. Phosgene *oxime* was developed for the military. As a liquid, it is yellow-brown, and as a solid it is colorless. It is far more lethal than the industrial variety.

**Odor:** Freshly mown hay, offensive.
**Symptoms onset:** Immediate
**Persistency:** Vapors settle in low areas
**Isolation distance:** 2000 feet
**Evacuation distance:** 4.6 miles
**Symptoms**

Immediate skin and eye pain and irritation of the nose and lungs. Affected areas of skin whiten, and red rings form around those areas quickly. Hives, itching, pain, runny nose, coughing and shortness of breath.

## Medical treatment of Blister Agent exposure

Immediate evacuation of the area *upwind*. Immediate full body decontamination. Airway assistance probable with large dose victims. Pain medication.

I cannot emphasize enough the importance of rapid skin decontamination and avoiding breathing blister agents. Long term health care may be necessary in any case.

# Riot Control Agents

Choking agents are substances mostly associated with domestic Police activities. There are 3 basic agents.

**CN** and **CS** are chemical compounds that represent *tear gas.*

These chemicals were intended to affect the eyes, but if inhaled, can cause plenty of discomfort. CN and CS begin working within seconds and last for ½ hour or more, depending on the amount of dosage. CN is the more powerful of the 2. **OC** is the third and newest variety, and its primary ingredient is *cayenne pepper.* Pepper spray is considered the safest agent. OC is also intended to temporarily blind the victim. All 3 of these agents can be delivered by small aerosol cans, or in a grenade like fashion. They are easy to conceal. The discharge of these agents in a confined space has the ability to cause mass panic which could result in injuries.

Decontamination is not readily necessary. Eyewash, or eye irrigation with water can help ease the discomfort, but no further medical treatment should be needed. Victims with emphysema or other breathing problems may experience prolonged discomfort, and may need medical treatment.

There have been several incidents lately in which personal defense pepper spray canisters have accidentally discharged on commercial flights, forcing the airplanes to return to the airport. TSA and other security personnel will not allow these items on airliners anymore. They can also have effects on passengers when discharged within checked baggage. If you feel a need to arrive at your destination with a personal pepper spray devise, be sure it is undamaged, that the safety mechanism is on, and wrap and seal it within a large zip-top plastic bag. Do not roll or crumple the bag, as it will need to expand should the devise detonate. This precaution will help *you* from becoming a terrorist!

(Be sure to check airline regulations on pepper sprays.)

# History of Biological warfare

Contrary to popular belief, the use of biological warfare did not originate during the "Cold War" years. Natural *Toxins, Viruses, Bacteria, and Fungi* have been present on our planet since the beginning of time. The earliest recorded use of biological warfare was by the Romans. By putting dead animals in the water supply of their enemies, they hoped to contaminate their drinking water with the rotting carcasses. In 184 A.D., Hannibal attempted to ravish his enemy by catapulting ceramic pots filled with poisonous snakes onto his enemie's ships. The Plague, or "Black Death" in 541 A.D. decimated the pop-ulation of Asia, Europe and Africa.(The plague was not a result of warfare, but occurred naturally) In 1346, the Mongols, invading  the Genoese port of Kaffa threw the bodies of Tartars infected with the plague over the walls into the city. Some medical historians consider this act to have caused the "Great Plague" of Europe. Biological warfare was used on American Indians in 1763, with sad effect during the French-Indian war. In order to diminish the strength and numbers of the Indians, blankets that had been previously used by smallpox victims were given as "gifts" to the Indians, resulting in millions of deaths. During World War I, Germans attempted to reduce our horses bound to the war in France by injecting them with glanders disease. Japan was the first to aggressively pursue the research of biological weapons when it formed "Unit 731". During the years between 1937 and 1945, it is believed that over 10,000 Chinese and Korean civilians and Allied POWs died as a result of their "experimentations". It is also believed by some that bubonic plague and other bacteria were deployed over parts of China. The attempts appear to have been ineffective.

The Nazi's began their efforts in WWII as well by conducting human experiments with various parasites and virus's. In 1943, Britain devised linseed cattle feed "cakes" infected with anthrax with the intention of dropping them from the sky over Germany. The U.S. Offensive Biological Weapons Program started in 1942, and was terminated in 1969. We never deployed biological weapons, but experimentation on dispersal methods took place within the United States. The experimentation was to calculate and monitor the movement of the agents particles throughout the air. Most experiments, such as the release of a harmless bacterium in Washington National Airport used inert substances. In San Francisco, however, an active agent was used. An organism called *Serratia marcescens* was released as an air-born spray. This incident is believed to have produced one death and infected10 people with a mysterious illness. Our biological stockpiles were destroyed in the early 1970s, but research for defensive purposes continues. The good thing about the American biological project is that facilities were minimal. The former Soviet Union had a lively campaign to develop and stockpile biological weapons. When they started the research and development of biologicals is unknown. In 1973 and 1974, the Soviet Politburo formed the "Biopreparat", a highly clandestine program to produce vast amounts of biological weapons. The Biopreparat employed as many as 50,000 scientists and technicians at its height. Many different strains of agents were developed, including anthrax, plague, and smallpox. These agents were stockpiled by the *hundreds of tons*. The Russian population did suffer from a 1979 release of anthrax from a manufacturing facility in the town of Sverdlovsk. This release produced 79 infections, killing 66 people. In 1984, America suffered it's first biological attack as a terrorist weapon in the Dalles, Oregon. This incident did not immediately present itself as an attack. Evidence of the illness's having been intentionally caused did not surface until epidemiologists discovered the true origin almost a year later. The agent used was *Salmonella typhimurium*. The motive can be considered political, as it was intended to sway the votes in a local election in order to gain control of the local government by sickening the opponent population on election day so they couldn't cast a vote. The perpetrators were members of the Bhagwan Rajneeshees cult. The bacteria was produced by a cult member, and other members secretly infected several salad bars in the area. 900 people became violently sick, but no deaths were reported. The use of *anthrax* as a terrorist's weapon became evident in 2001, when 5 letters containing varying strengths of anthrax took 5 lives and infected 2 dozen people in Florida and Washington D.C. A tremendous impact was felt due to these 5 letters. The Brentwood Post Office was closed for a long period of time for decontamination, and the whole U.S. Postal Service had to prepare for possible future attacks. This proved costly and demoralizing. As of January 2005, no one has been convicted of these attacks.

In the years following those events, hundreds of anthrax *hoaxes* have taken place throughout the nation, exhausting local fire and police departments, HAZMAT units and Federal resources. These hoaxes are taken seriously by responders. Each and every hoax must be investigated and the substance analyzed and identified. It is my belief that hoax perpetrators, when caught and convicted should be tried and sentenced as having performed an act of terrorism.

## Biological Agents as Weapons

The thought of anyone using biological agents is terrifying. Many of these agents can be developed in small, homemade laboratories, using easily obtainable materials. As you have seen in the biological history, the types of agents and the methods of dispersing them is endless. Bacteria's, viruses and toxins all exist in nature. They have previously been available via the internet for research purposes. Controls on the delivery of these "germs" have become stricter in the last few years, but it is possible that some have gotten into the wrong hands already. The possibility of these agents having been misdirected from the former Soviet Union also exists. Crude agents have the ability to cause epidemics, but to be a truly successful weapon, infecting large populations simultaneously, it must be "weaponized". The attack must also be coordinated with meteorological conditions as well. *The primary target for these germs is the lungs. Inhalation is the fastest, most effective way to contract these diseases.* Hence, they must be made into a "powder" form with particle sizes from 1 to 5 *microns* in size. This is the process necessary in order to *weaponize* the material. This process requires a very sophisticated laboratory with highly trained technicians to be successful. Biological agents, whether weaponized or not, are susceptible to degradation by light, temperature, humidity, and ultraviolet rays. All of these factors are in our favor. The costs associated with the production of a weaponized agent are expensive, and timing of the release must be near perfect for a successful large scale release. Unlike a chemical attack, explosives would not be likely as a delivery method for biologicals, as the heat from an explosion could kill the microbes. The ideal method would be as an aerosol.

An effective method of delivering the germs to a metropolitan area could be from a crop dusting airplane. Aerosol "generators" could also be strategically place around a city or in highly populated areas or buildings.(Emergency Departments Country-wide are urging large building supervisors to increase security around central heating and cooling systems, particularly in public buildings such as hospitals and Government facilities, as a release within the ducting could infect the occupants within the entire building.)  If the devise or agent is not discovered during or after a release, and all the necessary conditions prevail, the attack could prove successful. An attack like this will probably be totally invisible to its victims. Symptoms within the population may not present themselves for days…or weeks. Should it be a *contagious* disease and go unrecognized for a period of time, the incubated germs could be transferred a hundredfold to populations that were nowhere near the initial release.

The emerging infections and the possibility of an epidemic would most likely be discovered within local hospitals first. *The awareness of the signs and symptoms of these diseases by the medical professionals working in hospitals and clinics is extremely important in reducing the effects and transmittal of diseases.* The sooner that an outbreak can be discovered and reported, the better the chance that epidemiologists can track its source. The CDC has an elite group of medical professionals that do just that. These epidemiological "strike teams" are ready to pack up and head into the area in a moments notice. They are the Epidemic Intelligence Service. (EIS)

Another effective method would to be to release the agent in a mass transit system, using small aerosol cans or sprayers as seen in Japan in the 1995 attack, where members of the "Aum Shinrikyo" sect released sarin gas within the subway system, killing 12 and producing symptoms in more than 5,000 people.

An attack performed by vaccinated persons or terrorists is also quite feasible. The individual could slowly and methodically spread the germs that he/she was vaccinated against at will in many different cities and airports. A suicidal terrorist could cause a widespread pandemic if he/she spent lots of time on various airlines during the proper incubation time of the germ he/she had been exposed to. Should this person not yet visually show symptoms of the disease, or those symptoms go unnoticed by passengers and airline personnel, standard security detection and searches would find nothing exposing him as a terrorist.

Biological agents have been termed as "the poor man's weapon of mass destruction". Pound for pound, biological weapons can destroy more people than a nuclear bomb, and for a fraction of the cost. It would be unwise for a terrorist organization within the United States to release a biological weapon unless their members were vaccinated *prior* to the release. Subsequently, should a group outside the U.S. unleash the germs, it could backfire on them should the germ find its way back to their unvaccinated population. Let's hope that International terrorists are smart enough to consider that scenario before unleashing the fury of a pandemic on the world. Doctors and Epidemiologists around the world spent the better part of the 20th century eradicating the horrific smallpox virus from the world. The last recorded case of smallpox was in the late 1970s.

Humans are not the only targets for a biological attack. Infection of our crops, livestock, and the poultry industries could have a profound effect on our economy and food resources.

# Biological Agents

The primary biological agents of concern are the *virus, bacteria, and toxins.* There are many of each variety. For the purposes of this book, we will study some of the main strains that we are aware of, have been produced as weapons, and are of the greatest concern.

# Viral Diseases

Viruses are small particles. They cannot be seen without sophisticated microscopes and cannot be trapped with filters. They consist of a core of nucleic acid (RNA or DNA) surrounded by a protein shell. They can live outside, but can only reproduce *within* specific types of living cells.

Viruses cannot be treated with antibiotics, however antiviral drugs are effective against *some* of them, and many can be controlled by vaccinations. Some are *highly contagious,* but not all. Some examples of viruses

**One of many field hospitals for treatment of overwhelming numbers of victims of the "Purple Death" Flu pandemic of 1918.**

Photo courtesy of the National Archives

are: AIDS, West Nile, chicken pox, hepatitis, measles, mumps, polio, smallpox, yellow fever, common colds, shingles, rabies, genital and mouth herpes, and the common flu (influenza).

The flu virus has plagued the world for centuries. Many dismiss the flu as being like a common cold, but it kills tens of thousands each year. In 1918, the "Spanish Flu", aka "Purple Death" killed 675,000 people in the U.S. alone. This *pandemic,*(world wide epidemic) claimed the lives of over 40 *million* people before it ran its course.

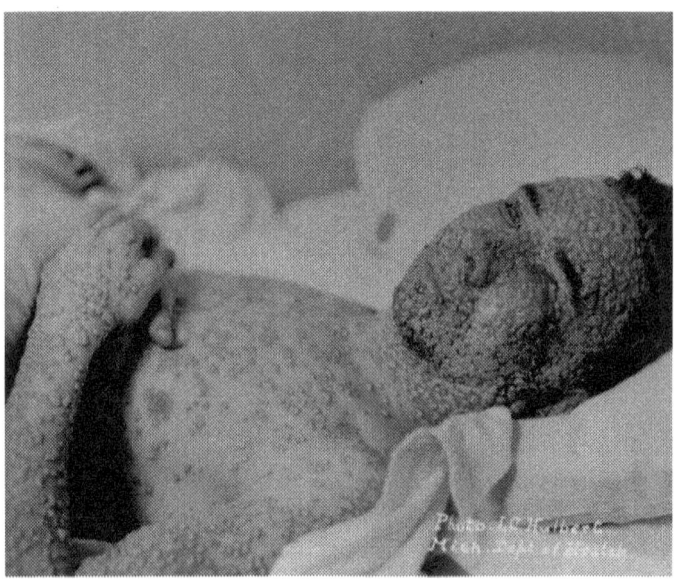

**Unvaccinated Smallpox victim. The virus proved fatal in this case.**

Photo courtesy of the National Archives.

# SMALLPOX VIRUS
*Variola major*

**Contagious:** HIGHLY CONTAGIOUS
**Route of entry:** Primarily inhalation. Also direct contact from fluids produced by smallpox lesions.
**Transmission:** Human to human
**Time from exposure to symptoms:** The incubation period can be anywhere from 7 to 17 days before symptoms arise.
**Initial symptoms:** fever, discomfort, headache, body aches, possible vomiting.
**Advanced Effects:** Red spots appear within the mouth. They then form open sores. Initially, it resembles chicken pox. Rash appears on the face, working its way down the body, arms, legs within one day. The rash then turns into bumps on the skin, which then turn into small raised blisters (pustules). The pustules then turn into scabs. After 3 weeks, the scabs will fall off. *The victim remains contagious until all of the scabs have fallen off.*
**Morbidity:** Permanent scarring of the skin at scabbed areas.
**Mortality:** 25+%
**Prophylaxis:** Vaccination with bifurcated needle
**Treatment:** Isolation!! Antiviral medications?

Note! Variola major is the more severe of 2 types of smallpox. Variola minor exhibits minimal mortality.

# VIRAL HEMORRHAGIC FEVERS

This group consists of many different varieties of the virus. Each have varying degrees of illness and symptoms. VHFs include Lassa, Rift Valley, Marburg, yellow fever, dengue, and others.

## EBOLA VIRUS
*Filoviridae*

**Contagious:** Yes.
**Route of entry:** Primarily skin, lungs possible.
**Transmission:** Human to human/animal to human. Direct contact with blood and secretions, bodily fluids from infected persons.
**Time from exposure to symptoms:** Incubation period 2-21 days.
**Initial symptoms:** Fever, rash, headache, sore throat, weakness followed by stomach pain, diarrhea and vomiting. Rash, red eyes and internal/external bleeding may occur.
**Advanced Effects:** Patients with advanced illness will experience hemorrhaging (bleeding) internally as well as externally.
**Morbidity:** Minimal in survivors.
**Mortality:** 50-90% of clinically ill.
**Prophylaxis:** None
**Treatment:** Isolation. No vaccines developed as of yet. Intravenous fluids and electrolytes. Monitor blood pressure.

# Bacterial Diseases

Bacteria are single celled microorganisms. They range in size and shape. *Cocci* bacteria are spherical, and range is size from 0.5 to 1 microns. *Baccili* bacteria are rod shaped and are 1 to 5 microns in size. Bacteria exists naturally in the soil, water, and in the air. The important aspect of bacteria as a weapon is that they *do not need a living host to survive, and some form spores that are resistant to ultraviolet light and heat.* Some bacteria, such as *anthrax* can be menacing due to these properties. Some bacterial diseases are highly contagious. *Antibiotics* can be used against most strains of bacteria effectively if administered soon after exposure and in sufficient doses.

Because of bacteria's diversities, more harmful strains were produced for warfare.

Some examples of bacteria with biological weapons potential and their natural means of infection are:

ANTHRAX, from grazing animals, also known to as "wool gatherers disease".

PLAGUE, from flea bites on rodents, aka "Black death".

TULAREMIA, from tick or fly bites infected by wild rabbits and rodents, aka "Rabbit fever".

BRUCELLOSIS, from livestock, aka "Undulant fever".

Q FEVER is also associated with livestock animals, aka "Query fever".

GLANDERS is associated with horses.

# ANTHRAX
*Bacillus anthracis*

**Contagious:** Not person to person
**Route of entry:** Inhalation, ingestion, skin (broken skin susceptible to entry)
**Time from exposure to symptoms:** *Inhalation*: 1 +days. *Ingestion:* 2-5 days. *skin*: (cutaneous) Up to 2 weeks.
**Initial symptoms:** *Inhalation*: flu like symptoms. *Ingestion*: nausea, diarrhea containing blood, then stomach pain.
*Skin:* Small sore that develops into blister, Blister turns into ulcer with black center. Pain not associated, but itchiness is. Fluids from ruptured ulcers *can* be transferred to others.
**Advanced effects:** *If not treated properly,* putrefactive destruction of tissue, hemorrhagic meningitis, death.
**Morbidity:** Possible scarring from blisters.
**Mortality:** *Inhalation:* 90-100%
*Ingestion:* 75-90%
*Skin:* 25%
Note: Mortality rates based on *untreated* victims.
**Prophylaxis:** Vaccination regimens.
**Treatment:** *IMMEDIATE oral* Antibiotics, *Ciproflaxin* effective. 2 month daily medication regimen.
*IMPORTANT! Flu symptoms may go away within a few days of inhalation exposure, but return with a lethal vengeance! If you suspect that you have been infected with anthrax, do not discontinue antibiotics!*

# PNEUMONIC PLAGUE
*Yersinia pestis*

**Contagious:** Highly Contagious
**Route of entry:** Primarily INHALATION
**Transmission:** Person to person (air-born)
**Time from exposure to symptoms:** 2-4 days
**Initial symptoms:** Pneumonia, chest pain, fever, coughing bloody sputum.
**Advanced effects:** Pneumonia persistent, leading to death.
**Morbidity:** None externally
**Mortality:** 50-60% if left untreated
**Prophylaxis:** Vaccinations available, but suggested only for high risk groups such as medical personnel.
**Treatment:** Antibiotics and support therapy (hospitalization)
Bubonic plague is another common form of plague, and is usually caused by infected flea bites. *Plague septicemia* can develop if not treated properly.
Bubonic plague is *not contagious*. Left untreated, it can have a 50-60% fatality rate.

# PNEUMONIC TULAREMIA
*F, tularensis type A*

**Contagious:** No
**Routes of entry:** Primarily inhalation when used as a weapon.
**Transmission:** Inhalation, ingestion, or skin absorption. In its "natural" state, transmission to humans is via insect bites and handling infected animals.

**Time from exposure to symptoms:** 3-5 days
**Initial symptoms:** fever, nausea, headache, joint and muscle pain.
**Advanced effects:** Pneumonia, inflamed and tender or painful lymph nodes. Death of cellular structures in tissue and organs.
**Morbidity:** Skin absorption may leave scarring.
**Mortality:** untreated, 5%   treated, 20-30%
**Prophylaxis:** Doxycycline
**Treatment:** Antibiotics 1-2 weeks.
There are other types of Tularemia. *Ulceroglandular Tularemia* causes skin ulcers and swollen lymph nodes, and is usually not fatal. *Typhoidal Tularemia* causes fever, cough, and respiratory infection, and fatal in 30-60% if untreated.

# Toxins

Toxins are poisons that occur in nature. The list of possible toxins is endless. A terrorist without the means to weaponize a toxin could find many toxins that are readily available as weapons. Some examples of toxins are: arsenic, castor beans (used to make *ricin*), cyanide, gasoline and oil products, certain mushrooms, leads, mercury, heroin, insecticides such as diazinon and malathion, and venoms from spider, ant, jellyfish, and snake bites.

Toxins are produced from living organisms, and are not as volatile as chemical weapons, or as deadly as the virus or bacteria's.

There are 2 toxins in particular that may be favorable for the advanced terrorist.

# BOTULINUM TOXIN
*Clostridium botulinum*

This is produced in an environment that is absent of air. It is one of the most potent toxins. Most associate this with food poisonings. It could be delivered to victims by ingestion or inhalation. *Symptoms occur 12-36 hours after exposure, and include headache, inability to swallow, and vision impairment. As it affects the central nervous system, muscle paralysis begins, moving from the head and down the body, impairing breathing when it reaches the lungs. Death occurs as suffocation.* The mortality rate is about 5%. Subsequent respiratory problems can continue to be problematic for years after exposure. Vaccines are available for prophylaxis, but must be taken *before* exposure to botulism. Antitoxins are available, and fatality can occur if left untreated.

## Ricin Toxin

Ricin is a by product of the *castor bean*. No, castor oil does not contain ricin, however ricin is used in the field of medicine. Castor beans grow in many parts of the world. It is a very toxic substance, hundreds of times more potent than VX by weight! Ricin works by attacking cells within the body. It causes inflammation of the stomach and intestines and the failure of other organs such as the liver and kidneys. Ricin is relatively easy to make, which is disturbing. It has been used as a clandestine weapon for assassinations by injection, but it wouldn't be prudent to think that terrorists would use this method. The most possible methods of exposure would be inhalation if aerosolized, or by contaminating our food supplies.

If ricin is delivered by means of the **food supply**; *nausea, stomach pain, and vomiting will be early symptoms; within a few hours. Bloody diarrhea follows soon after. Death from food-borne ricin could take place in 3 days or less.*

Should ricin be **aerosolized,** the symptoms would be different, as it will enter the lungs first. *Chest tightness, difficulty breathing, coughing, nausea and nasal congestion would be early symptoms, beginning 2 to 4 hours after inhalation. Lung problems will follow as they fill with fluids, restricting breathing. Death can take place anywhere from 36 to 72 hours.* There is *no antidote* for ricin poisoning. Treatment could vary from activated charcoal for gastronomical problems to assisted breathing. If you expect exposure to ricin, you should seek immediate medical attention.

# STRATEGIC NATIONAL STOCKPILE

The thought of biologicals being used in America is frightening, but we have been preparing for this eventuality for years. The Strategic National Stockpile, (SNS) is a prepackaged supply of medical equipment, supplies, and pharmaceuticals. (Formerly called the NPS, or National Pharmaceutical Stockpile.) There are many of these stockpiles located in logistically strategic cities. Their actual whereabouts is known only by those with Top Secret clearances. These huge caches of medicines are constantly maintained and highly guarded. Should an epidemic occur, local Health Departments can request delivery of a complete stockpile, or specific medicines, supplies and equipment needed to treat or provide prophylaxis to the population at risk.

Health Departments in many states are planning how these drugs can be distributed quickly and efficiently should they be needed. Many communities are having practice drills, or "exercises" to see how efficient their planning is, and to educate and prepare the public. Many different "models" are being developed, and not all communities will share the same methods of distributing the medicines. Small towns may only need to set up 1 or 2 clinics, while large metropolitan cities may need many in different sections of town. The planning for this type of an event is an enormous undertaking by your local Health Departments, Emergency Management Departments, EMS, hospitals and clinics. These agencies need to consider routes of traffic, parking, handicapped people, ethnic communication methods, signage, security, record keeping, public announcements and instructions, and many other aspects involved. Their intention is to establish and practice strategies in order to successfully distribute these medicines rapidly should the need arise. If your community should call for volunteers for these exercises, I strongly suggest you and your family participate. Most exercises only last a few hours, and the experience will be enlightening and educational. The cooperation of the population being vaccinated in a real life event is crucial to the success of the mission. The more you know about what to expect and where to go, the better your chances of survival. Should an actual epidemic occur, and you are notified of the clinics and given instructions on when and where to go for your medications, it is important that you follow the instructions. The distribution centers are being planned for a certain number of medications and medical providers. If you are told that your neighborhood will be vaccinated tomorrow, don't try to infiltrate another clinic today. This will only add to the stress of the care providers and their supplies. Should you need to go to a clinic for medications in a large event, you should wear your N95 facemask. It would also be a good idea to bandage any openings on the skin, such as cuts and abrasions. Long sleeve shirts and long pants will help as well. If your instructions are to come as a family unit, do so. You should also bring a ballpoint pen and drinking water. Epidemiologic questionnaires will need to be filled out by everyone. This is to gather information about the "path" of the disease, and to check for any health problems that may cause adverse reactions by the drugs being given. Be patient and calm. You can expect some people to be overly excited or panicked. Be cooperative. Due to the nature of the disease, and its progression of symptoms on different individuals, *you may not be able to stay together as a family.* It is possible that one or more members of the family will need advanced care in an isolation area. Be prepared for that possibility. Your calmness and cooperation will be of great importance in the success of getting medical care.

# NUCLEAR

We have all seen the devastating effects of a nuclear explosion in documentaries and movies about the first nuclear bombs unleashed on the cities of Hiroshima and Nagasaki in World War II. The destruction and injuries of a nuclear explosion are horrifying. It has previously been thought that the cost and technology would prohibit the procurement and use of nuclear weapons by terrorists, but that may not be true today. Terrorists may not be able to obtain and/or use a weapon as large as those used on Japan, but any nuclear explosion could produce a significantly disastrous event.

Nuclear events cause several types of destruction. *Blast & thermal* destruction are the primary types of destruction in the immediate area at the time of detonation. Thermal radiation from a nuclear blast causes 2 different types of burns on its victims. *Flash* burns are those created by the direct absorption of the infrared thermal energy on exposed surfaces. The other burns associated with the explosion are the *flame* burns, which are secondary fires set off by the explosion. Also resulting from a nuclear blast is *ionizing radiation*. There are different types of ionizing radiation. *Alpha particles* can partially be stopped by clothing, but areas that do receive this type of exposure can experience cellular damage. *Gamma rays* are also emitted during the explosion and are highly energetic, passing through matter easily. It is the gamma rays that cause the dangerous exposures resulting from "fallout". *Neutrons* are only emitted during the actual nuclear event and cause 20 times the amount of tissue damage than the gamma rays. Neutrons are not

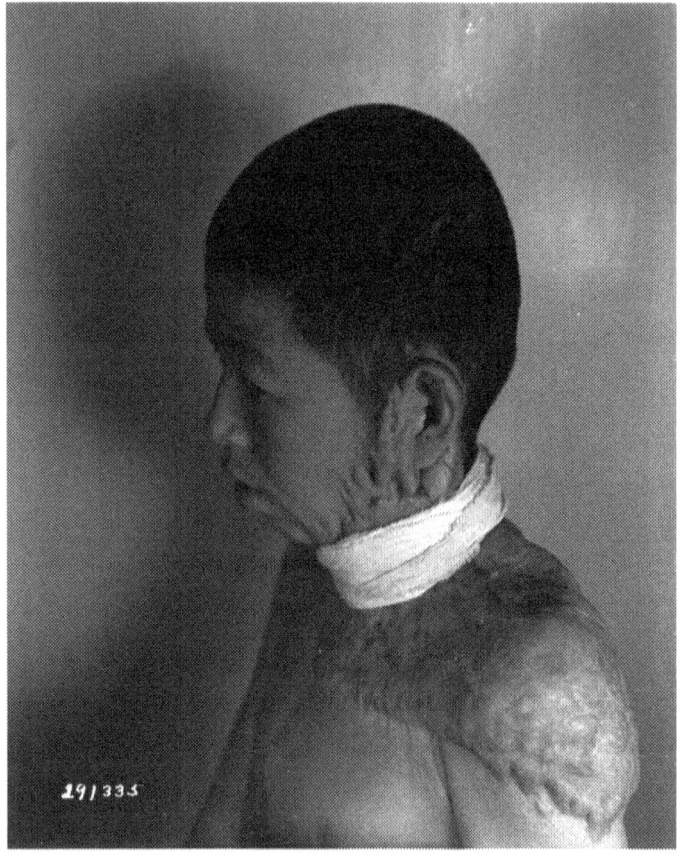

**Injuries from Nuclear explosion. Nagasaki, Japan.
World War II**

Photo courtesy of the National Archives

associated with fallout. Calculating the amount of exposure is based on the amount of radiation and the amount of time you were exposed to it.

People who work within nuclear facilities, test sites, or in industries where radiological elements are used must constantly track and record the amounts of radiation that they are exposed to. In order to do this, they wear *Dosimeter* cards. These credit card size devices are put on at the beginning of a shift, and turned in for readings afterwards.

Radiation dose rates are measured in *rad*, (radiation absorbed dose). A lethal dose, or "LD" for 50% of exposed humans is 450 rad, or *LD-50*. A new International radiation scale is slowly replacing the "rad" scale. It is the GY, or *Gray Scale* in which one GY=100 rad. Should a high or questionable amount of rad be discovered or suspected, the victim will be decontaminated and blood tests will be taken. The physician will be studying the *lymphocyte* counts, which are the blood cells most vulnerable to radiation.

Those lucky enough to have survived a nuclear blast or "dirty bomb" may not necessarily be out of harms way. Radiation can cause skin ulceration, eye injuries such as retinal burn, bone marrow invasion, gastrointestinal compromise, and lifetime or fatal cancerous ailments.

*A victim of radiation exposure may not have any initial symptoms.* Those people close enough to the blast will present obvious injuries such as blistered skin, vomiting, and difficulty breathing, or other obvious blast induced trauma.

Should you be in an area where a nuclear attack is imminent, and evacuation is out of the question, seek shelter *as far beneath the ground as possible.* Should there be a ventilation system or air source from above, turn it off and seal all openings to your shelter. Should you see a blast from a distance, try to put a building or other solid object between you and the blast to minimize ionizing radiation exposure.

# DIRTY BOMBS

Dirty bombs are the alternative way to cause radiation damage to a target. A dirty bomb is simply a conventional explosive containing radioactive materials. A dirty bomb is referred to as an RDD, or "radiation dispersing devise" by the military. These can be quite demoralizing and require extensive human and technological resources for a great deal of time after the explosion. The area in and around the explosion will need to be monitored for radiation, and "clean-up" of the area performed before the area can be repopulated.

It is certainly possible for a terrorist to get his hands on *some form* of radioactive material, as it is used in medicine and industry. *Those people working in the fields of medicine or industry where these materials are used and stored should keep a close eye on security and suspicious activity in and around the workplace.* Certain types of radioactive materials in the hands of Black Market dealers have been found by Counter Terrorist agents, but more may exist from stockpiles or "depleted" supplies from the former Soviet Union or other Nations. The United States has very strict regulations on the storage and transportation of radioactive materials, but it is not impossible to obtain them. Infiltration drills have taken place in the U.S. to examine the security of our radioactive material, and have shown that security measures need to be enhanced. With the heightened security at our ports and airports, as well as the sophisticated detection equipment employed by the security personnel, it would be difficult but not impossible for a terrorist to import a significant amount of radioactive material into the U.S.

# FALLOUT RADIATION DECONTAMINATION

Fallout radiation is composed of radioactive *particles*. Should you have been near an area subjected to, or suspected of a radiation release (nuclear or "dirty" bomb), these particles must be removed from your body. This is called the *decontamination process*, or "DECON". About 90% of the radioactive particles will be on your clothing or in your hair. To properly decontaminate, *gently and carefully* remove your clothing from the top down. Wash the hair with water while bending over forward, then proceed to thoroughly wash the entire body. *Radioactive particles can be transferred to another person.* The main difference between radiation decon and chemical decon is that radioactive particles are not liquids or vapors, and will "float" in the air if you are not careful while *gently* removing your clothes.

## MEDICAL TREATMENT FOR RADIATION

There is no specific immediate medical treatment for exposure to a nuclear blast with the exception of standard blast, burn and trauma treatments. There is, however a treatment for exposures from *Gamma rays* associated with fallout. The *thyroid gland* is the organ most susceptible to cancer damage from gamma rays, and damage can be minimized by immediate ingestion of *Potassium Iodate*. These tablets are available to the general public, and come in 65 or 85 mg doses. These *radiation blockers* should be taken for 3-14 days. Daily dosages are:

*newborn to 1 mo.*=1/4 tablet *1 mo. to 3 years*=1/2 tablet. *3-12 years*=1 tablet. *12 years and older*= 2 tablets.

# EXPLOSIVES

Explosives are the favorite weapon of the terrorist because of the ease in which materials can be found in industry, and the simplicity of the manufacturing process. There are many different substances that can be used. Timothy McVeigh's resourcefulness proved just that with the use of fertilizer (ANFO) in the destruction of the Federal Building in Oklahoma. Sophisticated explosive substances, while controlled, can still find their way into a terrorist's arsenal.

Some aspects that lie in favor of the terrorist are that the substances are readily available, a single person could carry out the attack remotely or, (as we see in the news almost daily) it can be delivered precisely to the intended target by a suicide bomber.

Internationally, 50% of terrorism involves bombings, while in America, 86% are bomb related.

In 1997 alone, Federal agents in the U.S. reported over 8,000 lbs of explosives, and over 3,000 detonators stolen.

I am enraged that some of *our own* military personnel are selling our weapons and ammunition on the Black Market. In May of 2005, two of our soldiers were caught trafficking weapons in Columbia. These weapons may have been intended for Columbian Paramilitary Groups (which in itself is a terrible thought), but they could very well have landed in the hands of terrorists, and destined for use in America!

One military weapon capable of intense destruction is the AT-4 (M-136) disposable, handheld rocket launcher. These rockets are less than 4 feet long, weigh under 5 lbs, and have an effective range of over ¼ mile. The precursor to the AT-4 is the "LAWS Rocket" (Light Anti-Tank Weapon). The LAWS rocket is the one used by Dirty Harry in the movie "The Enforcer", where he destroys a lookout tower on Alcatraz. Both the LAWS and AT-4 rockets are capable of penetrating 10 inches of armor, and can bring down airplanes, not to mention a wide variety of other targets.

Explosions release light, heat, and sound. Anyone that has been witness to a substantial explosion will never forget the power of those 3 properties.

The term "Energetic Materials" refers to 3 different types. *Explosives* release immediate energy. *Propellants* offer a controlled release of gas for a specific job. (Bullets and cannons are examples of propellants). *Pyrotechnics,* as in our 4th of July celebrations, produce heat, light, smoke and sound.

High explosives require a detonator of some sort, and larger explosive devises may need a detonator and secondary explosive to ignite the main explosion. Explosives can come in various forms. Nitro glycerin, which is an early form of high explosive, is a thick liquid. It is highly unstable, and many unintentional explosions were a result of its instability. Many miners and railroad workers used this in the 19th century to blast tunnels for the railroads. Those working with nitro glycerin experienced headaches constantly, not due to the explosions, but due to the vapors it omitted. One form of explosive, *Detasheet/flex-X,* which is ideal for letter or package bombs, comes in a flexible, thin sheet or roll. A rule of thumb to remember, is that if you don't know and trust the person you receive a package or letter is from, don't open it!

Some commonly used explosives used by the military are TNT, PETN, RDX, HMX, SEMTEX, and C-4.

The most common commercial explosive used today is Dynamite.

Terrorists and other bomb-makers may use ANFO, which is a mixture of fertilizer and fossil fuels. ANFO requires an initiating explosive of considerable power in order to be detonated.

All explosives require a type of detonator and a means to set the detonator off, (initiator). Homemade initiators can be devised using batteries, light bulbs, cigarette lighters, mouse traps, fuses, clocks, cell-phones, and many other various devices. Improvised explosive devices, (IEDs) all require some kind of detonator. These can be ignited by percussion, explosion, or electric shock. Most commercial and military detonators (blasting caps) have a unique look about them. They are generally small and resemble a tubular pen-like rod with wires coming out of one end. Some detonators will have long coils of wire attached to one end.

*As with radiological materials, those of you that work in environments that sell or use any of the components necessary to make an IED should be vigilant in paying attention to the sales, security and presence of these components. Should you suspect the misuse of these products, contact 911, or the FBI or ATF office in your area.*

The relationship between the *amount* of explosives being detonated, and the *distance* you are away from the explosion will dictate the amount of damage you may sustain.

When you suspect you are observing the placement of an explosive for terrorist purposes, you should get a safe distance away and call 911 immediately. The size of the explosive package,(briefcase, car, truck) will dictate the distance you should move away from that object. Putting a large object such as a building between you and the "package" can also increase your chances of survival should the devise be successfully detonated.

One thing to keep in mind is that a terrorist may detonate an explosive to draw in the public and Emergency Personnel, and then detonate a larger, more destructive device. These are called "secondary devices". The effectiveness of a *secondary device* was made quite evident in April 2005, when a bomb exploded in a Shiite Mosque in Baghdad. After the blast, people rushed in to help, and a suicide bomber drove into the area and detonated his bomb. The results were 15 dead and 57 wounded.

The explosive vests that are common to the walk-in suicide bombers contain more than explosives! There is also a layer of "shrapnel" covering the vest. This shrapnel can be nails, ball bearings, screws, rocks, or any other item handy that can cause penetration injuries as they fly through the air from the explosion.

*If you believe you are about to experience a suicide bomber (and cant get away,) get yourself behind a solid object (if possible) and in a fetal position on the floor, covering your head.* You may not be able to escape the blast, but can avoid being struck by the projectiles.

# GUIDELINES FOR SAFE DISTANCE FROM EXPLOSIVE DEVICES

| | |
|---|---|
| Small Box | 985 feet |
| Briefcase | 1,125 feet |
| Compact car | 1550 feet |
| Full size Sedan | 1800 feet |
| Van | 3800 feet |
| Box Van | 6500 feet |
| Large Commercial Trucks | 6500 feet |
| Semi Truck/Trailer | 7000 feet |

## Chapter Four

# DECONTAMINATION PROCEDURES

*What is a contaminant?* A contaminant is any substance that can cause you bodily harm when it comes into contact with your skin, eyes, ears, nose, and airway. Contaminants,(agents) can be liquids, vapors, powders, aerosols, or solids. Some industrial agents like chlorine, ammonia, gasoline or acids can begin irritating the victim immediately. Some may not show symptoms for hours or days later. If you are experiencing sensations from a contaminant whether it be an accident or act of terrorism, you will want to get decontaminated (DECON) as soon as possible.

Consider how you have been decontaminating yourself all of your life! What do you do if you spill gasoline or acid on yourself? If you spill it on your clothes, you will take the clothes off and wash the affected areas of your body with water and possibly soap, right? Those are the same basic principles that are used by the military, fire fighters and HAZMAT units around the world. One of the crucial elements in decontamination is moving *up-wind and uphill* from the spill or incident as quickly as possible and then perform the DECON process.

Your outer layer of skin is basically dead. Healthy skin will usually repel most irritating substances that come into contact with it for a period of time, but not for long! Remember, the most vulnerable parts of your body are the lungs, eyes, nose, and cuts and abrasions on the skin.

*In a terrorist incident where there is an agent released, it may take hours or longer for the emergency response teams to be able to identify what the agent is.* In an incident like this, you may need to perform your own self-decontamination as quickly as possible.

In most domestic industrial accidents, identification of the substance is rather quick, due to the heavily enforced HAZMAT substance signage, placards, and labeling requirements. These are the diamond shaped placards with a colored picture and numbers on them that you see on the back and sides of tankers (both highway and railroad). Emergency responders all carry what is called the "Emergency Response Guidebook", or ERG. By looking at the placard and referencing the ERG, the responders will know *what* the substance is, and *how* to proceed with response activities. The ERG also directs the responders as to how far away from the spill people should be moved or evacuated to. *During a terrorist attack, the terrorist will probably not let the nature of the agent be known in order to inflict as many casualties as possible.* Should this be the case, it may take many hours before the exact nature of the agent is known. Many Fire Departments have sophisticated chemical detectors that can sample and identify the agent shortly after entering the hot zone in level A protective suits, but those lacking the costly detectors can only *assume* the identity of the agent by observing the medical symptoms that the victims display. In a very large scale chemical (or biological) release, the Federal Government will deploy the nearest WMD-CST to survey and identify the agent, and to assist local responders in coordinating a successful response. The "Weapons of Mass Destruction Civilian Support Teams" are 21 person National Guard teams that specialize in the identification of chemical and biological agents. There are now over 25 such teams strategically located around the U.S. Their response time to an incident could be from 1 to 6 hours, depending on their base location. In the following chapter, "A citizen's role" I will go into some depth regarding the need for you to trust and *cooperate* with the first responders in terrorist acts and industrial accidents.

Should you suspect you have been contaminated, and need to perform self-decontamination, the rule of thumb is: *Get upwind of the incident, (and uphill if possible) get your clothes off, and wash your hair and body.* The importance of getting your clothes off is that they will probably contain most of the contaminants! DO NOT put those clothes back on! In an ideal situation, you will have access to plastic bags. If so, put all of your clothes, shoes, purse, wallet, *and jewelry* in a garbage bag and tie it up. The importance of including the removal of jewelry for the decontamination process is because contaminants may be trapped within the links or bands of watches, or under stones or rings. If you have 2 bags, you will want to put the jewelry and wallets in one bag, and the clothes in the other. The reason for separating non valuables such as clothing and shoes from valuables like jewelry and wallets is simple. First, if the agent is identified as an extremely harmful element, the clothes and shoes may be incinerated, as trying to decontaminate them and return them to the rightful owners would be time and cost prohibitive. Secondly, health officials may decide it prudent to decontaminate personal items and return them to their owners, but not willing to try and separate valuables from clothing. There are two reasons to try and put your possessions into sealed plastic bags. The most important reason is to try and contain the contaminants so they cannot spread to uncontaminated areas or people. Secondly, retrieval of your items will be easier should health officials deem the agent inert.

There are 2 different scenarios that are most likely to happen that will require DECON. The first is where you have the ability to DECON yourself. Let's say the attack happened in a shopping mall, and you had your preparedness kit *and* a gallon or 2 of water in your car. Providing you can get to your vehicle, you could DECON yourself in a reasonably modest way in a short period of time.

Should you do this, it is best that you remember to stay upwind in your closed vehicle and turn on the emergency radio channel for further information after you have decontaminated yourself. Any source of water will work, but in highly populated buildings and malls, it may be difficult to find a spigot or restroom, and if you do, there will probably be a lot of people ahead of you! Should the exposure be of a quick working, highly irritating nature, panic and selfishness will surely be present. Panic is the one ingredient that will make *any* emergency dangerous and hard to control. The next time you are in a mall or just walking around your city, keep an eye out for water spigots. You will probably find few that are readily available, or for public use!

The second scenario reflects activities that will take place after Emergency Services arrive. Should it be assumed or proven to be an act of terrorism using WMD, you may not be allowed to leave the scene. There are 2 reasons for this. The first is that the whole area will be considered a *crime scene,* and the perpetrator and/or evidence may still be in the area. The second reason is that contaminated people cannot be let loose to contaminate others outside the incident. The area will be secured by law enforcement, and *zones* will be established and marked. The sizes and locations of these zones will depend on the size of the contaminated area and the wind direction. The *Hot zone* is the area that has the most contamination.( The area that the agent was dispersed in.) The *Warm zone* is an area upwind of the hot zone that has a degree of safety. While waiting for decontamination, you will need to remain in a designated area within the warm zone. The *Cold zone* is the safe area upwind from the warm zone. *You will not be able to cross into the cold zone until you have been decontaminated!* Once you have been decontaminated and go into the cold zone, you cannot cross back into the warm zone for any reason, or you will once again need to be decontaminated. Here is where you must be sympathetic towards the response personnel trying to save lives, and be considerate for others around you! Proper decontamination is a time consuming process, and having to DECON the same person twice will lessen the chances for other peoples survival while adding undue stress to the DECON team. You must remember that responders cannot work in their level A or B suits for more than one hour (or less, depending on conditions), before they must rest, re-supply their oxygen, and check the integrity of their equipment. This is why DECON units consist of two or more teams. While one team is working, the other is resting and refitting. You must also remember that *ambulances will not transport you, and you will probably not get medical help until you have been properly decontaminated.* Attempts by contaminated people to enter medical facilities will only compromise the medical personnel, the facility, and the patients present at the hospital or clinic. If a medical facility has been contaminated, it could cause a temporary or long term closure of the facility. After Emergency Services (usually the fire fighters) recognizes the need for the immediate decontamination of the victims, there are many different ways they might do it. Many fire departments are obtaining sophisticated DECON units, but most don't have the budgets to procure such items. (The Department of Homeland Security is making funds available for these types of items to emergency agencies and hospitals). Some DECON units are collapsible Quonset hut shaped corridors. Some are inflatable, and some are actually built into DECON trucks. All of these types of DECON units are designed to be able to wash as many people in as short of time, and *with as much dignity and privacy as is possible.*

Another method, if specialized DECON units are not available, could be to simply provide "rain" with fire hoses that people could wash themselves with as they walk through. Remember that first responders must make the best use of any and all resources within their reach to save lives, but might not be able to provide the privacy. All responders will want to DECON the victims as thoroughly as possible. Unfortunately this does require the removal of clothes and jewelry. A DECON team may make you go to a *guarded* isolation area if you refuse to cooperate with proper DECON protocols. *Those that do cooperate will be the first to be decontaminated and receive proper medical attention.* Many fire departments now carry disposable gowns or other garments to wear after decontamination. There are also many "property protection" packages being developed for your personal items such as jewelry, purses and wallets. Some systems are bar coded, and others may be as simple as zip lock bags that your name or social security number are put onto with a felt pen. These systems are being developed to assure you that your items will be returned to you either after the items themselves are properly decontaminated, or it is proven that the agent released presents no harm. You must remember that *no contaminated items will be allowed into the cold zone.* The property protection packages will need to remain in the warm zone with your discarded clothes before you can proceed into the decontamination corridor.

The whole response and DECON process is difficult for the first responders, but that difficulty is compounded by the inability to communicate personally with the people they are treating. The people operating the DECON corridor operations (in the warm zone) will be wearing PPE. (personal protective equipment) These are encapsulating "space" suits. They need them to assure that they don't become casualties as well. If they get exposed, then they can't help others.

These suits are bulky, hot, and claustrophobic. They make it almost impossible to talk to or hear others, especially when there is peripheral noise. You may get instructions from pamphlets or flyers, signs, loudspeaker, or repeating recordings. *Preparing for communications in areas with diverse ethnic populations is a daunting task.*

One factor that is disturbing when considering the indignities and inconveniences experienced in *proper* decontamination is the possibility of the event either being a *hoax,* or containing non-hazardous agents. Many will *prefer to believe* that it is a hoax, and refuse to participate in the decontamination process. Those individuals may laugh at those that cooperated should it indeed have been a hoax, but on the other hand, they might end up sorry or dead for their stubbornness or modesty.

The need for a thorough and proper decontamination of victims must not be ignored! *One droplet of nerve gas smaller than the size of a pin head is enough to kill several people.*

There are several different additives for the water to make more effective decontamination solutions. The military prefers a 0.5% Sodium hypochlorite (household bleach) solution. Most household bleaches are 5% pure. Mix 1 part bleach to 10 parts water. HAZMAT units have other specialized products, but a bleach solution works fine for most exposures. The bleach solution is best for Biological agents but may cause irritation of cuts, wounds, and rashes due to chemical exposure. Hand or body soap will help to decontaminate as well. Also good is grease cutting dish-soap. Just remember, that the most important ingredient is *water!* The military also has personal decontamination kits that they carry with themselves while operating in the field.

The older version is the M258A1, which may be available at war surplus stores. It is about the size of a can of spam, and comes with saturated wipes and papers. It is intended for decontaminating smaller exposed areas like the hands and face. The newer version is the M291.

Whether you are decontaminating by yourself, or being decontaminated in a corridor, you will still need to follow the same basic procedures.

**1) Get upwind and uphill**

**2) Remove all clothing and jewelry**

**3) Wash thoroughly.**

Start by tilting your head forward and washing the hair so the water falls on the ground instead of running down your body. Wash and rinse *every inch* of your body as quickly and thoroughly as possible. Blow your nose. Pay particular attention in cleansing the ears, nose, mouth, and any cuts or abrasions on your skin. Wash the feet last, then step into the cold zone and clean surface. Whenever possible try to stand on something that is above the water level at your feet while decontaminating.

*When being decontaminated by response units, wash as quickly as possible in consideration of those waiting in line behind you. Do not be surprised or angry if you are required to repeat the process again, or later at the hospital or medical aid station.*

**4) Do not go back into the warm zone!**

Don't jeopardize yourself after finally making it to safety by going back into a contaminated area or by putting your contaminated jewelry or clothes back on. Don't try to retrieve your personal belongings until they have been proven to be safe by health officials.

## Chapter Five

# A CITIZENS ROLE IN COUNTER-TERRORISM

Hopefully, after being introduced to the terrorist's and the emergency responder's roles and methods, you will now have a better idea of how *you* can help prevent and respond to an act of terrorism. Your knowledge is one of your best weapons, but only if you *use* that knowledge. As demonstrated on the Southwest Airline flight to Florida on January 25, 2005, civilians are stepping up to the plate to *physically* stop hijackings and other terrorist like attempts. Some passengers on planes on 9/11 also attempted to stop the terrorists. These are exceptional individuals that are willing to risk their personal safety. The more people that can help these individuals, the better chance they will have for success. Many things can be used as both defensive and offensive weapons on airplanes. Laptops, seat cushions, and books are some examples. Create as much confusion for the terrorist as possible! The chances of actually being present and able to take these measures during a terrorist attack is slim. Most of us can help the fight against terrorism in many other ways that don't require bravery and fighting skills.

## Preparedness

Having your family's preparedness plans and supplies in a constant state of good order is possibly the best thing you can do. By doing this, you can assure you will have the ability for self sufficiency which will take a great deal of pressure off of the first responders and those working in the emergency management department, while increasing your chances of survival. *Remember, it may take days or longer to receive aid from local or Federal sources.*

## First aid

The more skills you have in first aid, the more you can help yourself, your family, and others. First aid courses are available in all communities, and are not expensive. A basic first aid and CPR class is usually 4 to 8 hours long. Make the class a family affair. All will benefit from it whether or not the skills are used in a terrorist attack. Refresher courses, intended to keep those skills honed are also available. These are courses that only take 2-4 hours, and are primarily a quick overview of the ABCs of first aid as well as CPR. A refresher course should be taken every 2 years. Contact your local Office of Emergency Management or the local Red Cross to find out where and when these classes are available.

# Volunteer

A very important thing to remember is that large scale incidents will be inaccessible to citizens without *identification of affiliation with a response group* working within or around the disaster site. Security for these incidents has escalated since 9/11. The intentions of "volunteers" are admirable, but often their efforts are counter productive, and put themselves and trained responders in harms way unless they are trained and working with a response group. Large scale incidents, particularly bombings, and anything that may be a criminal or terrorist attack are regarded by law enforcement, fire fighters, and other first responders as *Crime Scenes.* Not only are these officials trying to save lives, but they are also trying to detect and preserve any evidence. The evidence recovered may lead agencies to important clues that can prevent future incidents.

There are many volunteer groups that play significant roles in large scale disasters. The members of these groups are given the appropriate training, and once capable of performing the duties required, are presented with identification that will allow them into the area to help those in need. The volunteer experience is rewarding. You will meet and have the ability to work and train with great people. Some volunteer agencies are more active than others. The Red Cross often responds to local emergencies such as house or wildfires; feeding, sheltering, and providing mental assistance to responders and victims alike. The Red Cross volunteers were called upon during and post 9/11 to help the responders and families of the victims. Their role was instrumental in helping to maintain order. Another example of a volunteer agency is the National

Disaster Medical System. This group, a response division of FEMA, is comprised of over 50 specialized teams, located throughout the Nation that are called into action on Presidentially declared disasters where their particular skills are needed. These teams consist of doctors, nurses, paramedics, psychologists, logisticians, communications technicians, Morticians and Veterinarians.

Another excellent agency is the Citizens Corp, which has many opportunities to help the local communities. There are thousands of these programs throughout the Nation. There are also Senior, and Medical Reserve Corps. All of these groups are also great volunteer agencies for retired people to become involved with as well.

There are many other volunteer groups that may need someone with your special skills. The Civil Air Patrol could use pilots and radio operators. Ham radio clubs often play an important role during disasters, providing communications when phone lines are down. Volunteer agencies can always use anyone with special skills such as computer operators, food handlers, security specialists, logisticians and laborers.

To find out which volunteer agencies exist in your area, contact your local department of emergency management, or go online.

The important thing is to get involved as a volunteer *now!* Attempts to volunteer during or shortly after a disaster will prove fruitless, as the agencies are busy directing their existing volunteers and there is no time for training, background checks, and other administrative tasks necessary in making you a member. During 9/11 and in the wake of the 2004 Indonesian Tsunamis, as the representative for the DMAT HI-1 team, I fielded countless calls from people who wanted to join and respond to those incidents. These were concerned people with the best intentions. I had to give them the distressing news that they would not be able to respond to those incidents. I did urge them to go ahead and begin the application process so they could be used in the next disaster. Sadly, after sending out almost 50 application packets to these individuals, only 3 people actually completed and submitted them. However, those 3 people now have completed the required training programs and have the proper identification in order to participate in a future event.

Many people are reluctant to join a volunteer organization because they "don't have time". Most volunteer outfits are sensitive to the time that their volunteers have to spend with the group, and will try to accommodate individuals whenever possible. Naturally, the more time volunteers spend interacting as a unit, the better response capability that group will have. Before joining as a volunteer, it is best to take the time and discuss it with your family and get their blessing. Then, when enrolling in your volunteer group, be up front from the very start as to how much time you can spend training and attending meetings.

*Don't wait…*Locate the organization that best fits your skills and *become a volunteer now.*

## Awareness and reporting suspicious activities

Your awareness and willingness to report suspicious activities is of immense importance to all response agencies.

People working in all of the different trades need to keep a watchful eye on the going's on in their workplaces. Sports stores and gun shops must be more cautious than ever when selling guns and ammunition to unknown individuals. Companies that sell or launder uniforms should keep track of inventories and be assured their customers are actually legitimate, licensed professionals.

This also applies to imports! Millions of tons of imports enter our country every day. In May of 2005, Federal Agents confiscated 1300 well made metal I.D. badges, representing 35 different Law Enforcement agencies. These badges, coming into San Francisco, had they gotten into the wrong hands, could have been a great asset for the terrorist.

All industries must pay careful attention to the security and destination of any products that a terrorist could use for his mission. A terrorist may be in the market for bomb making supplies such as pipes and chemicals from hardware stores, items used for detonators from electronic stores, and fuel oil or fertilizer used in the production of the explosive "ANFO". When making sales to unknown clients, should you suspect foul play, try to develop a friendly conversation with him in order to get useful information, such as; "Wow! That's a whole lot of fertilizer! What kind of plants do you grow? Where is your farm?" Most honest people won't have a problem with being friendly, but too many questions will probably soon irritate the bad guys. If you remain suspicious, try to find a reason to see his I.D. If not, try to record his vehicle type and license plate number. In doing your own "investigation" of the purchaser, you have just taken the first step in the counter terrorism process.

Now you need to remain focused and move on to step two…Reporting suspicious activity.

Reporting your concerns and observations is easy. When in doubt, call 911. The most important thing to remember when making these reports is to stay calm. The last thing a dispatch operator needs is to have to try and understand a hysterical caller. Here are some guidelines for calling in any emergency or suspicious activities.

**1)** Take a deep breath and organize your thoughts!

**2)** Speak slowly and calmly.

**3)** State your name and location.( If you are a member of a response or volunteer organization, let it be known. This may help the operator to believe your report is credibile.)

**4)** Present your report in an orderly fashion. Be as accurate as possible about times, locations, etc.

**5)** Be prepared to give the operator a location or phone number where you can be reached.

## BOMB THREATS

Should you receive a bomb threat by telephone, there are a number of things to remember to do *during* the phone call. First of all, remain *calm*. If possible, record the call! Try to listen for any background noise. This may help in locating the suspect. Write if possible, but do your best to remember the entire conversation. Pay attention to speech patterns, such as a shaky voice, loud talking or screaming, language dialects, whispering, etc. Any of these can give investigators important clues. Whispering, for example might mean that the caller is in an area where people might be able to hear him. The slightest detail might help prevent the event and/or apprehend the caller. Try to talk to the caller calmly. Don't call them names like "sicko" or "nutcase". This could make them even madder. Ask them *why* they are doing this. Ask them *where* the bomb is if he hasn't told you already.

Ask him where he is. If you have caller ID, remember to write the number down before the caller hangs up. If you don't have caller ID, try to retrieve the number using your local phone recall service *before* making another call. Many phone companies use *69 to retrieve the last number dialed to you. Call 911 to report any bomb threat, and follow the procedures suggested for any 911 call. Do not try calling the ATF, as they are investigators, not responders. Should the ATF need to be pulled in on the investigation, local law enforcement officials will request them.

## Photography

There is nothing more true than the phrase "A picture is worth a thousand words". A video can speak volumes.

Should you be in the area of an incident, (or see suspicious behavior or materials) and have camera or video equipment with you, use it! Most disaster photo's and news clips show the misery and pain. That's how the media makes a living. Your focus should be on the people that are *in and around* the area. From where you are standing, try to get pictures of the people 360 degrees around you. Just keep clicking. Often the perpetrator will stick around to see his handiwork and the chaos it brings. If the bombing or agent release required the perpetrator to detonate the devise close to it, your photo's might contain him leaving the scene. Any photos of an incident during or shortly afterwards are valuable to investigators. Get the film to the FBI or police *immediately*. Don't try to get them developed first. Federal and local agents have the uncanny ability to expedite photo processing! Digital cameras and cell-phone cameras are wonderful tools, as they can be downloaded and "mailed" to the FBI instantly. *Do not* manipulate the images in any way!

# Medical Facilities

During a disaster producing large numbers of trauma patients, all local clinics, hospitals, and ambulance services will be overwhelmed. Most hospitals try to limit their staff to the normal needs for the community. When disaster strikes, all of their off duty doctors and nurses will return to work to help balance the health care capabilities within their facilities. Panic and confusion can cause serious congestion on roads leading to the medical facilities. Whenever possible, if your injuries are minimal, don't add to the confusion if you don't need immediate treatment. If you are able to walk, you probably don't need to go to the emergency room, and if you do make it into the ER, you will be *triaged* and might end up in a waiting line.

Triage is the process where a group of casualties or patients are "sorted" according to the seriousness of their injuries so that the available medical technicians can prioritize which injuries should be treated first. *Triage, in emergency situations, is designed to save as many lives as possible.*

Another problem that hospitals experience is the "worried wounded". These are the people that *think* they might be in medical trouble, but aren't. This phenomenon is most likely to occur during an incident where a chemical or biological agent was used. This occurred during the Tokyo subway Sarin gas release, and kept the medical community very busy for days after.

## Emergency Communications

Any time there is a large scale emergency, the possibility of the phone system becoming overwhelmed exists. Make *only* emergency related calls in these situations, and make the calls as *short* as possible, focusing your discussion only on *necessary*, pertinent information. Your cooperation will be appreciated by, and give rescue units and emergency operation centers a greater chance of success.

## Personal Protection Equipment

This is a subject that comes up quite often. You may remember the scramble for plastic sheeting, duct tape and gas masks at the time of the Anthrax incidents. Isolating a room from the outside environment is a valid procedure, but don't forget you need air! Hospitals have *isolation rooms* for patients with contagious disease or need protection from the outside environment. These rooms, in order to be safe and effective are *positive pressure* areas. The force of the clean air being generated or pumped into the room provides oxygen for life, and forces the outside air from coming into the room. Before the air is exhausted to the outside, it is cleaned, or "scrubbed" with HEPA filters.

Gas masks are dangerous! Most military gas masks were developed for short operations, and contain about 250 ml of "dead space", which means that it does not purify outside air for long. It is not recommended to rely on these for any extended period of time, as you may asphyxiate yourself.

Sophisticated chemical resistant suits and breathing apparatus of varying degrees of protection are available to the public, but have 2 drawbacks for civilians. The firemen, HAZMAT, and CST personnel that work in these are in *exceptionally excellent physical condition*. People with medical ailments such as high blood pressure and breathing disorders, or that are claustrophobic and not in excellent physical health should not consider using them.

The other thing that the professionals have is proper and continuous *training*. These suits need to be *fit tested* every time they train with them to assure there are no holes in the suits, and that the face and respiration equipment is working properly and all seals are not leaking.

Should you desire to have some form of PPE, I suggest purchasing TYVEK suits and N95 facemasks. These are relatively inexpensive and available at most safety supply stores in your area. The TYVEK suits are lightweight, white suits with booties and a hood. They are resistant to entry of small particles, but *not* chemical resistant. The TYVEK suit can protect a good deal of your skin, but leaves exposed areas on your face and hands. Rubber boots and rubber gloves can be added to the TYVEC suit for more protection.

Before purchasing any PPE, consult with your doctor. Also study the specifications and limitations of the products before purchasing or relying on them. *Remember, any PPE is only as good as the amount of time you practice wearing and maintaining it!*

# First Responders-Emergency Management

First Responders are here to *save lives.* Their first priority is to save the lives of people, then property, and then the environment. Emergency management officials have the job of coordinating all of the resources needed by the responders in order for them to do their work. All of these people are highly trained in their disciplines and will do their best to accomplish their missions, no matter how small or large the incident is. The thing they need most from you is *respect and cooperation.* Try to pay attention to their needs. One thing they absolutely need is room on the road to drive their emergency vehicles! Pay attention while you are driving your vehicle. When you hear a siren or see the lights, don't wait until it gets close to you. Try to find out where it is coming from, and pull over as necessary. When you pull over, you might start the chain reaction that EMS and Fire Fighters needs most. They can't help anyone if they cant get to the scene!

You must remember that first responders, while saving lives and property are also trying to be sure unnecessary *collateral damage* is not done. In these days of terrorism, the First Responders have to consider a wide range of issues before jumping into a potentially dangerous situation that previously would not have been necessary.

The terrorist seems to enjoy harming and killing first responders. It upsets the local response infrastructure and demoralizes the first response communities. This is the reason they need room to move around and within the incident area… and your total cooperation. You may want to help, and it's *possible* that you will be allowed to help in extreme situations, but probably not. The first responder may not let you in the area for fear of a *secondary device.* These are bombs or traps intended for the first responders. Remember, their job is to protect and save lives.

Every day, while either training or responding to emergencies, the EMS, Fire Fighters, and Law Enforcement Officials are putting themselves in harms way in order to insure our safety. What can we do for them?

One thing we can do for our local Emergency Departments is to vote! Pay attention to local and regional legislation and budget issues that can help the response community to fund manpower, equipment, and the necessary training. Be sure to carefully study the issues before voting though, as you may be voting against the good guys!

## Responsible Journalism

Throughout my career in emergency management, I have had the opportunity to work with the media with varying degrees of success. In one instance, I requested that the location of our WMD training facility not be disclosed. That newspaper writer and her photographer respected the need for the team's security, and did not disclose our location, but got a good story and great photographs anyway. In another instance regarding a news-paper, I strongly advised against over sensationalizing our team's capabilities, as it would give the community a false sense of hope. I thought I had gotten through to her, but the headlines the next day proved me wrong. Many emergency departments are wise and develop a good understanding and relationship with the newspapers and television stations (and visa-versa) before an incident takes place. In such a way, reporters know what details are acceptable to report, and what should not be reported due to safety and security issues. A *responsible* journalist must be aware of the dangers that can result from inaccurate reporting and the disclosure of operationally sensitive information. They must remember that the terrorist is watching and reading too!

As I began writing this book, I wondered if I could actually pull it off. Why? I knew too much about operationally sensitive issues. I wanted to educate the general public, but not provide the terrorist with any ammunition that could compromise the safety of our responses. I wanted to provide the public with as much information as possible to help them be able to protect themselves, but not to generate undue fear or sensationalism.

It is fine to sell newspapers and newscasts, but *don't sell out our Nation's security!!*

## Stress Management

The stress associated with massive trauma and destruction can leave psychological scars that can stay with you for a lifetime. Some cannot be prevented, but the effects can surely be lessened by psychologists trained in *Critical Incident Stress Management.* Many people in the first response communities take courses on CISM, and this training has proven to be invaluable to those having witnessed massive trauma, pain and suffering. Knowing how to deal with stress and trauma *before* you are exposed to it can make all the difference in the world as to the quality of your life *after* the incident. Also included in stress management is the "Critical Incident Stress Debriefing". This takes place at the end of the responder's shift, before they go home. The debriefing session may take form as a group session, or individually. Some responders are traumatized to the extent that they need further counseling.

Dealing with this kind of stress should not be taken lightly. Those of you that are involved with emergency operations are urged to take courses on CISM. It is also important that the wives and husbands of emergency workers have a basic education of stress management so they can be aware and sensitive to the needs of their partners when they return home. This basic education can also help in recognizing latent signs of stress related to the disaster.

If you are not a member of the response community, and lack CISM training, I suggest you take an hour and search some websites. There are a lot of sites that will give you valuable CISM information for free. Should you or a family member witness a highly traumatic experience, you should consider contacting a CISM specialist to ensure that you or aren't subjected to a lifetime of nightmares.

## International Travel

While the main focus of this book has been dedicated to terrorism on American soil, I do feel it necessary to discuss safety issues for those that travel outside the United States. Now that kidnapping and suicide bombings have gained popularity among terrorists, travelers will need to exercise more caution than ever before. Following some simple guidelines could help prevent you from becoming a victim. Remember that when a kidnapping takes place outside the U.S., the incident is controlled by the host Government. American response to these incidents can be frustrating and time consuming, if there is any response at all! *As an American citizen, while traveling abroad, you must remember that you, as an individual representing our Country, will not compromise with Terrorism, and can expect little or no help if you put yourself in harms way.*

The first line of defense is to *research the area you plan to visit.* The more you know about the political climate, the language, money system, and the people, the better off you will be. You should also study the geography.

Before leaving the country, you should make *black & white* copies of your Identification, Visa, and passport. Keep these hidden well in the event you lose the originals or they are stolen. It is also a good idea to write your passport number down in your wallet. *You should present your passport only when absolutely necessary!* For your security, it is better that you remain anonymous as much as possible. If you are traveling with someone else, you should exchange copies of photo identification with them, so each will have the means to identify the other to officials in an emergency situation. *Never copy identification in color*, as, if stolen, it could be used as I.D. for the bad guys!!! You should also plan to keep documents and money in separate pockets. Keep small bills in a pocket separate from large bills.

*Try not to look like an American!!* Remember that the terrorists don't like Americans or their affluence! Don't wear expensive clothes, shoes, or jewelry. Those new air filled "Nike" shoes are a dead give away that you are an American. Try to meld into the local crowd as much as possible. This also pertains to the amount and type of luggage that you are traveling with. Travel as light as possible, and don't use fancy luggage. Old and mis-matched luggage is recommended.

*Learn the language*, or as much as possible before getting on the plane. Learning how to ask directions, locate a bathroom, and order a meal in the native language will help in maintaining your anonymity as an American. Becoming familiar with the money and product pricing prior to entering the country will also help you to appear more "local". Try to speak English as little as possible.

*Choose your hotels and transportation wisely!* Many of us travel to experience the feeling and meet the peoples of the land we are visiting. Traveling, after all, is an adventure!!

Being a hostage, however is not my idea of adventure. Try to select a hotel run by a large, reputable company. Also, try to get a room on the lower floors, but avoid the bottom floor. You will find a certain degree of security staying in an American operated hotel where the employees speak English. They can arrange cabs, suggest restaurants and shopping, and advise you of areas that should be avoided. As soon as you check in to your hotel room, you should familiarize yourself with your surroundings. Locate fire extinguishers and evacuation stairways and assure yourself that you can find them in the dark.

*Avoid routines* as much as possible. Take different routes to the markets and restaurants, and vary the times you go. Those tourists that have predictable habits will be the easiest targets for kidnapping.

*Wear an air of humble confidence.* If you conduct yourself as if you are in control, you may be passed over as an "easy target" by the terrorist. Americans are known to be aggressive and often arrogant to the peoples of third world or developing countries. Presenting yourself with a bit of humbleness can gain favor from the locals, as well as shield your identity as an American.

*Be responsible for your own security.* Be cautious about each and every decision you make. Be extra cautious about things that you would not normally concern yourself with. Don't walk next to buildings when crossing alleyways.

Avoid darkened streets and parks. Travel with a partner whenever possible. Inspect your hotel room and make sure the windows and doors lock and open properly. When selecting a rental car, make sure it runs good, but pick one that will not stand out as a tourist car. Inspect the car for strange wiring or odd looking devices. Look under the hood as well as under the vehicle itself. Look under the car each time you go to get in, and if you notice anything odd or suspicious… don't get in! Keep your car doors locked while it is parked, and while you are driving. Use common sense…

It is unfortunate that we cannot enjoy the freedom and excitement of traveling to mystical lands as we have before. It is equally unfortunate that we feel the need to disguise ourselves as "other than Americans", but we must take any and all precautions to ensure our safety while traveling abroad, and when necessary, even postpone our travels until our destinations are deemed safe from terrorism.

# Conclusion

America's war on terrorism is an enormous task that will surely take years, decades, or longer to win. Terrorism has gotten so popular world wide that its flames may never die out. Our biggest enemy, Al Qaeda and its followers have proven their distaste for Americans, and their willingness to take extreme measures to hurt us. They don't like apple pie and their hatred for Americans runs deep… The capture of Bin Laden will not assure us of the end of terrorism. Other Muslim groups with the human and financial resources and willingness to die for their cause will still present a threat for years to come. In our efforts to "stabilize" the Mid East, we seem to have "stirred up a hornets nest". And remember, there are other fanatical groups throughout the world and in the U.S. that are also a great concern to our security.

We must do our best to protect our own soil. We must be diligent in protecting our sea and air ports, borders, railways, factories, cities, homes, agriculture…and people. It is a daunting task for Law Enforcement to accomplish. They need our help. Every American needs to join in the fight against terrorism. We need to think "out of the box" and be observant of our surroundings. We need to report suspicious activities in a timely manner. Our efforts in personal and family emergency preparedness are crucial. The many excellent volunteer organizations together can be a powerful force in responding to acts of terrorism. Working as a whole; American citizens *can* make a difference in the fight against terror.

I hope you have gained a better understanding of the hazards, how to recognize them, and how to respond to disasters and terrorism when it strikes. And most of all, I hope you become proactive and join the fight for the safety of our Nation, our freedom, and the American way of life.

# Acronyms

| | |
|---|---|
| **ATF** | Bureau of Alcohol, Tobacco and Firearms |
| **BCW** | Biological Weapons Convention |
| **BST** | Burn Specialty Teams |
| **CBRNE** | Chemical, biological, radiological, nuclear, explosive |
| **CD** | Civil Defense Agency |
| **CDC** | Center for Disease Control |
| **CISM** | Critical Incident Stress Management |
| **CP** | Command Post |
| **CST** | Civilian Support Team (National Guard WMD teams) |
| **CW** | Chemical, Biological |
| **DEA** | Drug Enforcement Agency |
| **DECON** | Decontamination process |
| **DEM** | Department of Emergency Management |
| **DHS** | Department of Homeland Security |
| **DMAT** | Disaster Medical Assistance Team |
| **DMORT** | Disaster Mortuary Teams |
| **DOD** | Department of Defense |
| **DOT** | Department of Transportation |
| **EIS** | Epidemic Investigation System |
| **EMS** | Emergency Medical Services |
| **EOC** | Emergency Operations Center |
| **EOD** | Explosive Ordinance Division |
| **EOP** | Emergency Operations Plan |
| **FBI** | Federal Bureau of Investigation |
| **FEMA** | Federal Emergency Management Agency |
| **FRP** | Federal Response Plan |
| **HAZMAT** | Hazardous Materials |

| | |
|---|---|
| **ICS** | Incident Command System |
| **JTTF** | Joint Terrorism Task Force |
| **MMST** | Metropolitan Medical Strike Team |
| **NDMS** | National Disaster Medical System |
| **NTSB** | National Transportation Safety Board |
| **PPE** | Personal Protection Equipment |
| **USAR** | Urban Search and Rescue |
| **USAMRIID** | U.S. Army Medical Research Institute for Infectious Disease |
| **VMAT** | Veterinary Medical Assistance Team |
| **WHO** | World Health Organization |
| **WMD** | Weapons of Mass Destruction |
| **WTC** | World Trade Center |

# Glossary

**Aerosol:** A suspension of insoluble particles in a gas or vapor.

**Agent:** The active substance in a chemical or biological weapon.

**Ambulatory:** Patients that are able to walk.

**Bacteria:** Single celled microorganisms.

**Cold Zone:** The area outside of the warm zone of a chemical or biological release that is deemed safe and clear of contaminants.

**Contagious:** A disease that spreads from person to person.

**Dirty Bomb:** A conventional explosive that contains chemical or radiological agents.

**Epidemic:** Rapid, widespread growth of a disease.

**Epidemiology:** The study and geographical tracking of a disease.

**Fallout:** The descent of radiological particles (gamma rays) to the earth following a nuclear event.

**Hemorrhagic:** Conditions in the body associated with extreme blood loss.

**Host:** A person or organism in which a parasitic organism lives.

**Hot Zone:** The dangerous area within and around a chemical or biological release.

**Incident:** Responder terminology for an accident, disaster, or crime scene.

**Incendiary:** Bombs or devices that are designed to start fires.

**Immunization:** (Inoculation) Using vaccines to prevent disease.

**Inert:** A substance with no active ingredient (agent).

**Irritant:** A chemical which causes inflammation of tissues.

**Lacrimation:** The production of tears in the eyes.

**Micron:** A particle the size of one millionth of a meter.

**Morbidity:** The state of being diseased.

**Mortality:** The death of the living.

**Neurotoxin:** Nerve Agent which is poisonous or harmful to the nervous system.

**Pandemic:** An epidemic so widespread that it spreads to vast numbers of people in different parts of the world.

**Parasitic:** Fungi, bacteria, virus, and worms that partially live within a living host.

**Persistency:** The ability for an agent to remain present for a considerable length of time after it's release. Usually pertains to chemical agents that are oily or sticky.

**Proliferation:** To rapidly increase the amount of a substance.

**Prophylaxis:** Any means taken to prevent a disease.

**Secondary Devise:** A bomb or agent placed on or near an incident, intended to harm the responders.

**Sludge:** Chem-Bio term for the release of copious amounts of bodily fluids. (Vomiting, runny nose, diarrhea)

**Toxin:** Poisons that occur in nature, or are produced using those poisons.

**Triage:** The act of sorting casualties (types of injuries) in order of their seriousness. Emergency triage is designed to maximize the number of survivors.

**Vector:** An animal or insect that transmits disease. (Rats, mosquitoes, ticks)

**Vesicant:** Any agent that causes blistering of the tissue.

**Virus:** Microscopic particles that can live outside of, but can only reproduce within specific types of living cells.

**Warm Zone:** The area surrounding the hot zone that is not entirely safe from an agent's effects.

**Weaponize:** The act of refining chemical or biological agents into their purest form. This includes making the particles small and light. Weaponizing a biological agent can also include coatings on the particles which allows them to withstand the environment for longer periods of time.

# About the Author

Ronald G. Laes was a Coastguardsman in the Vietnam era.

He is a founding member of the Department of Homeland Security.

As Administrative and Training Officer for the HI-1 Disaster Medical Assistance Team, (a division of the National Disaster Medical System and the U.S. Public Health Service) since 1998, he has had extensive training in the operational aspects of emergency response to disasters and Weapons of Mass Destruction. Mr. Laes has been the liaison for DMAT HI-1, assisting in emergency planning for natural and industrial disasters, WMD incidents, and bio-terrorism with State and County Emergency and Health agencies in Hawaii. On September 11, 2001, he was preparing a WMD Medical Strike team for training in Hawaii.

Ronald lives with his family in Washington state. He continues to work within the Disaster Preparedness and Emergency Response community.